Project Management for Performance Improvement Teams

The LITTLE BIG BOOK Series

Other books in THE **LITTLE BIG BOOK** SERIES

Project Management for Performance Improvement Teams

Authored by

William S. Ruggles and H. James Harrington

CRC Press
Taylor & Francis Group
Boca Raton London New York

CRC Press is an imprint of the
Taylor & Francis Group, an **informa** business

A PRODUCTIVITY PRESS BOOK

CRC Press
Taylor & Francis Group
60 Broken Sound Parkway NW, Suite 3
Boca Raton, FL 33487-2742

© 2018 by Taylor & Francis Group, LLC
Productivity Press is an imprint of Taylor & Francis Group, an Informa business

No claim to original U.S. Government works

Printed on acid-free paper

International Standard Book Number-13: 978-1-138-49715-3 (Hardback)
International Standard Book Number-13: 978-1-4665-7255-3 (Paperback)

This book contains information obtained from authentic and highly regarded sources. Reasonable efforts have been made to publish reliable data and information, but the author and publisher cannot assume responsibility for the validity of all materials or the consequences of their use. The authors and publishers have attempted to trace the copyright holders of all material reproduced in this publication and apologize to copyright holders if permission to publish in this form has not been obtained. If any copyright material has not been acknowledged please write and let us know so we may rectify in any future reprint.

Except as permitted under U.S. Copyright Law, no part of this book may be reprinted, reproduced, transmitted, or utilized in any form by any electronic, mechanical, or other means, now known or hereafter invented, including photocopying, microfilming, and recording, or in any information storage or retrieval system, without written permission from the publishers.

For permission to photocopy or use material electronically from this work, please access www.copyright.com (http://www.copyright.com/) or contact the Copyright Clearance Center, Inc. (CCC), 222 Rosewood Drive, Danvers, MA 01923, 978-750-8400. CCC is a not-for-profit organization that provides licenses and registration for a variety of users. For organizations that have been granted a photocopy license by the CCC, a separate system of payment has been arranged.

Trademark Notice: Product or corporate names may be trademarks or registered trademarks, and are used only for identification and explanation without intent to infringe.

Visit the Taylor & Francis Web site at
http://www.taylorandfrancis.com

and the Productivity Press site at
http://www.ProductivityPress.com

I dedicate this book, once again, to my wife and business partner,

Violeta Mercedes Granja de Ruggles ("Vivi Ruggles"). You have

been an inspiration to me for the past 39 years and the reason

for any successes I've had in my adult life. ¡Tu éres mi vida!

William S. Ruggles

I dedicate this book to Joe Cronin, who helped me become the person I am.

H. James Harrington

Contents

Preface

Welcome to the latest addition to the Little Big Book series: *Project Management for Performance Improvement Teams* (PM4PITs).

In another book in this series entitled *Effective Portfolio Management Systems*, the co-authors wrote that "one of the biggest organizational wastes occurring today in both private and public organizations is the high percentage of failed projects and programs." A "failed" project/program is one that does not meet one or more of its value-added objectives in terms of scope, quality, schedule, budget, or risk. While that earlier book in the series focuses on how Portfolio Leaders and PMO Directors can "do the right projects" in the strategic context of an organizational project portfolio, this book lays out how Project and Program Managers and their teams can "do those right projects the right way," **one project at a time**. If this is what you are looking for, you have found the ideal book in the ideal series!

Before delving into a summary of the scalable, innovative, and practical ideas you'll encounter in this book, there are four other books you should know about since they are **directly** related to much of the content covered in this one, providing in-depth treatment of four related topics: Performance Acceleration Management, Value Propositions, Business Cases, and Organizational Portfolio Management Systems. Let's take a brief look at what each one of those four books focuses on.

First, there is *Performance Acceleration Management (PAM): Rapid Improvement to Your Key Performance Drivers* by H. James Harrington, which is also the first book to be published in this series. Dr. Harrington found that organizations around the world have been rating their improvement efforts as failing to produce the desired long-term results. His research indicates that this occurs because organizations are NOT using the latest improvement tools and Approaches properly. They are trying to wield them without first defining **how they want to change their organization's culture, environment, and key performance drivers**. Proof-positive that a "Fool with a tool is still just a fool!" since 90% of a successful initiative is based on human factors, not technology! (Authors' note: Yet, having said that, we will cover Project Technology Management in Chapter 4 of this book.)

FIGURE P.1
Performance Acceleration Management book cover.

Organizations must first define which controllable factors drive business results: the key performance drivers. Then, they must define how they want to change these key performance drivers and behavioral patterns. Only then can they select a customized set of tools and approaches that will bring about the desired transformation.

The next book in the series is *Maximizing Value Propositions to Increase Project Success Rates* by H. James Harrington and Brett Trusko. These co-authors found that a Value Proposition, while being an old concept, is taking on new significance in today's innovation-driven environment. Business focus has shifted from developing many creative ideas to developing only those that will successfully flow through the product cycle and fulfill a customer need. This requires a crystal-clear value proposition.

FIGURE P.2
Maximizing Value Propositions book cover.

The third pertinent book is *Making the Case for Change: Using Effective Business Cases to Minimize Project and Innovation Failures* by Christopher F. Voehl, H. James Harrington, and Frank Voehl. It provides executive teams and change agents with the information required to develop better Business Cases and make better project selection decisions. This book illustrates how to develop a strong Business Case that links project and program investments to results and, ultimately, with the strategic outcomes of the organization. In addition, this book provides a template and exemplary Case Studies for those seeking to fast-track the development of a Business Case within their organization.

The fourth book in the series that is relevant to this one is *Effective Portfolio Management Systems* by Christopher F. Voehl, H. James

FIGURE P.3
Making the Case for Change book cover.

Harrington, and William S. Ruggles. It provides a roadmap for the implementation of Organizational Portfolio Management (OPM) using a Project Portfolio Management (PPM) system as a model for effectively driving sustainable change. It takes you through the complete portfolio management life cycle—from the submittal of the individual proposed projects, to their selection, deferment, or rejection, and, for those that have been selected, to the management of their implementation.

Now, having shared these four related resources with you, let's focus on what to expect from THIS book: *Project Management for Performance Improvement Teams* (PM4PITs).

This book provides straightforward guidance in plain English for Project Teams—especially Performance Improvement Teams (PITs)—and their Project Managers on how to successfully complete individual projects

FIGURE P.4
Effective Portfolio Management Systems book cover.

and programs using an ingenious and scalable framework based on an innovative foundation fusing together elements of Project Management, Innovation Management, and Continual Improvement.

This book's Introduction focuses on what continual improvement, change, and innovation are, why they are so important, and how they apply to performance improvement—both incremental and transformative. There, we take a closer look at the four types of work and workforce management in organizations, *Strategic, Operations, Projects,* and *Crises,* using four common comparative variables: Proactive/Preventive versus Reactive/Corrective, Temporary/Unique versus Ongoing/Repetitive, Innovative versus Maintaining the Status Quo, and Schedule Focus: Fiscal Year versus Short Term versus Long Term.

These comparisons set the stage for the uniqueness of the third type: **Projects** (and Programs) that are fundamentally change-driven. We also introduce two Project Case Study Examples—one for a nonprofit healthcare grant program with a portfolio of dozens of small projects and the other for a for-profit biopharmaceutical company with a single, large project.

In Chapter 1 we identify opportunities for improving each framework by pointing out three challenges or shortcomings with the "traditional" framework for Project Management and four challenges or shortcomings with the "traditional" framework for Continual Improvement. As a result of these shortcomings, neither one of these frameworks has been meeting the demands being placed on them in today's projects and programs, which require a more iterative and adaptive, yet scalable Approach.

In Chapter 2, we introduce, as upgrades, contemporary frameworks for both Project Management and Continual Improvement that address the shortcomings identified in Chapter 1. Then, we fuse them together to create a new combined framework for managing projects and programs that functions in a truly iterative and, if necessary, adaptive fashion to carry out performance improvement projects in a scalable framework with TWO Approaches: a "Full" Approach and a "Lean" Approach, depending on each project's relative Scope and Priority. (*Author's note*: By "Lean," we mean "a greatly scaled-down version of our 'Full' approach for projects whose Scope is relatively small and/or Priority relatively low." Figure 5.4 illustrates this concept using two distinct graphics.)

We identify 12 Project Management Performance Domains: **Integration, Scope, Schedule, Cost, Quality, Resources, Communications, Risk, Procurement, Stakeholders, Change,** and **Technology**, that traverse through five iterative Stages: **Align, Plan, Execute, Check/Act On,** and **Confirm**, each of which is also covered in more detail in Chapters 5 through 9. We describe how our contemporary framework is a truly iterative model to manage performance improvement projects and programs.

(Authors' note: Since we embrace the 49 processes as they are described in Chapters 4 through 13 in the sixth edition of the *PMBOK® Guide*, we see NO need to repeat those descriptions in this book. Hence, we urge the reader to obtain a copy of this 2017 publication and refer to it when needed. As such, we only provide in-depth coverage of the two new Performance Domains—**Project Change Management** and **Project**

Technology Management—and their 12 processes in Chapters 3 and 4, respectively.)

In Chapter 3, since resistance to change occurs on all strategic and continual improvement initiatives, we provide you with a "roadmap for driving change" comprised of the six iterative processes of the Project Change Management Performance Domains: **Align for Change, Enroll for Change, Plan Change Management, Manage Change, Check Change Management**, and **Confirm Change**. In this chapter, we want to help you overcome the inevitable resistance to change that will likely surface throughout your project's life cycle and the campaign to apply our Iterative Framework for managing performance improvement projects, via the use of a Change Management Plan. We close this chapter by continuing to share how its contents apply to one of the two Case Studies mentioned earlier.

In Chapter 4, since technology has become so ubiquitous in project management, innovation management, and continual improvement initiatives, we focus on Project Technology Management (not to be confused with "planning or managing a technology project") which is also comprised of six iterative processes: **Plan Technology Management, Implement Technology, Document Technology, Train on Technology, Report on Technology**, and **Integrate Technology**. This "Performance Domain" addresses the hardware/infrastructure and software resources needed to successfully carry out the performance improvement project. We close this chapter by relating how its contents apply to one of the two Case Studies first mentioned in the Introduction.

Chapter 5 is focused exclusively on the first Stage of our new iterative Framework for Performance Improvement: **Align the Project** (placing or arranging things in a straight line; providing consistent support to a given purpose or cause.)

Alignment is crucial to the successful fulfillment of the performance improvement project's goals since it ensures alignment of the project with its Business Case and Benefits Management Plan. While this Stage does NOT have many potential processes to perform, we still provide both "Full" and "Lean" Approaches for completing it properly, depending on the project's relative Scope and Priority. We close this chapter by relating how its contents apply to one of the two Case Studies mentioned earlier.

Chapter 6 is dedicated to describing the second Stage of our scalable Framework for Performance Improvement: **Plan the Project**. Since this Stage has A LOT OF potential processes to perform, we provide both

"Full" and "Lean" Approaches for completing it properly, depending on the project's relative Scope and Priority. Then, we close this chapter by relating how its contents apply to one of the two Case Studies mentioned earlier.

Chapter 7 is dedicated to describing the third Stage of our new iterative Framework for Performance Improvement Projects: **Execute the Project Work**. Since this Stage also has several potential processes to perform, we provide both "Full" and "Lean" Approaches for completing it properly, depending on the project's relative Scope and Priority. We close this chapter by relating how its contents apply to one of the two Case Studies mentioned earlier.

Chapter 8 is dedicated to describing the fourth Stage of our new iterative Framework for Performance Improvement: **Check/Act On the Latest Performance Data**. Since this Stage also has several potential processes to perform, we provide both "Full" and "Lean" Approaches for completing it properly, depending on the project's relative Scope and Priority. We close this chapter by relating how its contents apply to one of the two Case Studies mentioned earlier.

Chapter 9 is dedicated to describing the fifth and final Stage of our new iterative Framework for Performance Improvement: **Confirm the Results to Date (and Iterate?)**. The primary output from this realm is the Alignment Confirmation where you verify or confirm that you've successful fulfilled the performance improvement project's goals and are delivering the benefits that were promised in its Project Business Case and Project Benefits Management Plan. Even though this Stage only has a couple of potential processes to perform, we still provide both "Full" and "Lean" Approaches for completing it properly, depending on the project's relative Scope and Priority. We close this chapter by relating how its contents apply to one of the two Case Studies mentioned earlier.

Chapter 10 describes the importance of sustaining the gains and realizing the benefits that were promised in the latest version of your Project's Business Case and Benefits Management Plan and verified during the "Confirm the Results to Date" Stage via the Alignment Confirmation Plan.

Chapter 11, most of which is covered in the *Effective Portfolio Management Systems* book in this Little Big Book series, briefly reminds you of the importance of the Project Management Office and aligning your performance improvement project or program with its strategic direction at all times.

In the Epilogue, we share, just before submitting the final manuscript to the Publisher, our closing thoughts about the scalable, iterative Framework presented in this book and how it can be adopted by YOUR organization.

Finally, please don't overlook the Glossary of Terms and the Index in the back of this book. They should help you address any confusion you might encounter with the terminology along the way.

We hope you enjoy and benefit from this book by seeing a measurable 20% improvement (or MORE) in the success of your performance improvement projects! Please let us know how you do!

- William S. ("Bill") Ruggles: bill.ruggles@ruggles2llc.com or (201) 956-7905
- H. James ("Jim") Harrington: hjh@svinet.com or (408) 358-2476

Acknowledgments

I would like to acknowledge all of the clients I have had over the past 30 years. I hope that the practical experience I gained while working as a consultant and subject matter expert in project management and continual improvement for you is reflected in this book. I greatly appreciate the faith you had in me, my professionalism and conscientious approach to help you overcome your performance st*ruggles*!

I would also like to acknowledge Dr. H. James Harrington, who invited me to be his co-author of this book and, then, let me take the lead in bringing it to fruition.

William S. Ruggles

About the Authors

William S. "Bill" Ruggles is the Chief Operating Officer of Ruggles (aka Ruggles & Ruggles, LLC), a niche management consulting firm based in the New York City Metro Area. He is an internationally recognized expert in managing projects, programs, and portfolios and leading performance improvement teams with over 30 years of experience.

PM4PITs is his third book with the previous two being: *Effective Portfolio Management Systems* (2015) and the *Project Workbench Whiz's Sourcebook* (1990). He has also authored many articles and blogs under the pen name *Dr. Will B. Struggles.*

He has provided consultative, coaching, mentoring, training, and auditing support services to over 250 organizations and some 25,000 participants in 21 countries on four continents in two languages. These include many of the world's largest organizations in the fields of healthcare, biopharmaceuticals, biotechnology, medical devices, information technology, global communication technologies, insurance, banking/financial services, high tech manufacturing, engineering, shared services, and the public sector for both the state and federal governments. He was an Adjunct Professor of Project Management at Stevens Institute of Technology for 12 years and has also lectured at William Paterson University, New Jersey Institute of Technology, Essex County College, ITESM/Monterrey Tech, Universidad del Valle, ESPOL de Guayaquil, Universidad San Francisco de Quito, Indonesian PT Telekom RISTI Center, Fundación Chile, Turkish Makro Education Center, Boston University, and UTStarcom University.

His practice specialties include delivering quality and continuous process improvement outcomes, and enhancing project and program performances via the development of organizational, interpersonal, and individual proficiencies and competencies among those who sponsor, manage, and work on project, program, and portfolio teams.

Bill was certified in 1987 as a Project Management Professional (PMP #133) by the Project Management Institute, and has been a Certified Quality

Manager (CQM), a Certified Six Sigma Master Black Belt (CSSMBB), and a Certified Scrum Master (CSM). He holds an MA from Columbia University and a BS from the University of Connecticut and has performed MPH studies at the Icahn School of Medicine. He is fluent in English and Spanish.

Ruggles served as the president of the Project Management Institute's New Jersey Chapter (1990–1992) and served on PMI's International Board of Directors as its vice president of Technical Activities and president of the PMI Educational Foundation (1994–1995), as an ex officio member of the Board of Directors (1996), as president and chief operating officer (1997), and board chair (1998). He was also a recipient of both the Distinguished Contribution Award and the Linn Stuckenbruck Person of the Year Award in 1996. More recently, he served as the vice president of Administration for PMI's New York City Chapter (2011–2012) and as an advisor for its New Jersey Chapter (2013–2014).

Finally, Bill is a member of the American Society for Quality (ASQ) and has served as a Baldrige Framework-based *Examiner* at the State ("Quality New Jersey" in 2005) and National ("BNQA" in 2016) levels, and as a Baldrige Framework-based *Monitor* at the Regional ("Mid-Atlantic Alliance for Performance Excellence" in 2017–2018) level.

Dr. H. James Harrington is one of the world's quality system gurus with more than 60 years of experience. In the book, *Tech Trending*, Dr. Harrington was referred to as "the quintessential tech trender." The *New York Times* referred to him as having a "...knack for synthesis and an open mind about packaging his knowledge and experience in new ways—characteristics that may matter more as prerequisites for new-economy success than technical wizardry...." He has been involved in developing quality management systems in Europe, South America, North America, Middle East, Africa, and Asia.

PRESENT RESPONSIBILITIES

Dr. Harrington now serves as the Chief Executive Officer of Harrington Management Systems. In addition he serves as the President of the

Altshuller Institute. He also serves as the Chairman of the Board for a number of businesses and as the U.S. Chair on Technologies for Project Management at the University of Quebec in Montreal. Dr. Harrington is recognized as one of the world leaders in applying performance improvement methodologies to business processes.

PREVIOUS EXPERIENCE

In February 2002 Dr. Harrington retired as the COO of Systemcorp A. L. G., the leading supplier of knowledge management and project management software solutions. Prior to this, he served as a Principal and one of the leaders in the Process Innovation Group at Ernst & Young. Dr. Harrington was with IBM for over 30 years as a Senior Engineer and Project Manager.

Dr. Harrington is past Chairman and past President of the prestigious International Academy for Quality and of the American Society for Quality Control. He is also an active member of the Global Knowledge Economics Council.

CREDENTIALS

The Harrington/Ishikawa Medal presented yearly by the Asia Pacific Quality Organization was named after Dr. Harrington to recognize his many contributions to the region. In 1997, the Quebec Society for Quality (MQQ) named their quality award "The Harrington/Néron Medal" honoring Dr. Harrington for his many contributions to the quality movement in Canada. In 2000 Sri Lanka's national quality award was named after him. The Middle East and Europe Best Quality Thesis Award was named "The Harrington Best TQM Thesis Award." The University of Sudan has established a "Harrington Excellence Chair" to study methodologies to improve organizational performance. The Chinese government presented him with the Magnolia Award for his major contribution to improving the quality of Chinese products.

Dr. Harrington's contributions to performance improvement around the world have brought him many honors and awards, including the

Edwards Medal, the Lancaster Medal, ASQ's Distinguished Service Medal, and many others. He was appointed the honorary advisor to the China Quality Control Association, and he was elected to the Singapore Productivity Hall of Fame in 1990. He has been named lifetime honorary President of the Asia Pacific Quality Organization and honorary Director of the Asociación Chilena de Control de Calidad.

Dr. Harrington has been elected a Fellow of the British Quality Control Organization and the American Society for Quality Control. He was also elected an honorary member of the quality societies in Taiwan, Argentina, Brazil, Colombia, and Singapore. He is also listed in "Who's Who Worldwide" and "Men of Distinction Worldwide." He has presented hundreds of papers on performance improvement and organizational management structure at the local, state, national, and international levels.

Dr. Harrington has two Doctor of Philosophy Degrees: one in Quality Engineering and honorary PhD in Quality Management.

Dr. Harrington is a very prolific author, publishing hundreds of technical reports and magazine articles. He has authored or co-authored over 55 books, and 10 software packages.

The following is a list of some of the many books that Harrington has authored and co-authored.

- *The Improvement Process* (1987)—one of 1987 best-selling business books
- *Poor-Quality Cost* (1987)
- *Excellence—The IBM Way* (1988)
- *The Quality/Profit Connection* (1988)
- *Business Process Improvement* (1991)—the first book on Process Redesign
- *The Mouse Story* (1991)
- *Of Tails and Teams* (1994)
- *Total Improvement Management* (1995)
- *High Performance Benchmarking* (1996)
- *The Complete Benchmarking Workbook* (1996)
- *ISO 9000 and Beyond* (1996)
- *The Business Process Improvement Workbook* (1997)
- *The Creativity Toolkit: Provoking Creativity in Individuals and Organizations* (1998)
- *Statistical Analysis Simplified: The Easy-to-Understand Guide to SPC and Data Analysis* (1998)

- *Area Activity Analysis: Aligning Work Activities and Measurements to Enhance Business Performance* (1998)
- *ISO 9000 Quality Management System Design: Optimal Design Rules for Documentation, Implementation, and System Effectiveness* (1998, co-author)
- *Reliability Simplified: Going Beyond Quality to Keep Customers for Life* (1999)
- *ISO 14000 Implementation: Upgrading Your EMS Effectively* (1999)
- *Performance Improvement Methods: Fighting the War on Waste* (1999)
- *Simulation Modeling Methods: An Interactive Guide to Results-Based Decision Making* (2000)
- *Project Change Management: Applying Change Management to Improvement Projects* (2000)
- *The E-Business Project Manager* (2002)
- *Process Management Excellence: The Art of Excelling in Process Management* (2005)
- *Project Management Excellence: The Art of Excelling in Project Management* (2005)
- *Change Management Excellence: The Art of Excelling in Change Management* (2005)
- *Knowledge Management Excellence: The Art of Excelling in Knowledge Management* (2005)
- *Resource Management Excellence: The Art of Excelling in Resource Management* (2005)
- *Six Sigma Statistics Simplified* (2006)
- *Improving Healthcare Quality and Cost with Six Sigma* (2006)
- *Making Teams Hum* (2007)
- *Advanced Performance Improvement Approaches: Waging the War on Waste II* (2007)
- *Six Sigma Green Belt Workbook* (2008)
- *Six Sigma Yellow Belt Workbook* (2008)
- *FAST: Fast-Action Solution Technique* (2008)
- *Strategic Performance Improvement Approaches: Waging the War on Waste III* (2008)
- *Corporate Governance: From Small to Mid-Sized Organizations* (2009)
- *Streamlined Process Improvement* (2011)
- *The Organizational Alignment Handbook: A Catalyst for Performance Acceleration* (2011)

- *The Organizational Master Plan Handbook: A Catalyst for Performance Planning and Results* (2012)
- *Performance Accelerated Management (PAM): Rapid Improvement to Your Key Performance Drivers* (2013)
- *Closing the Communication Gap: An Effective Method for Achieving Desired Results* (2013)
- *Lean Six Sigma Black Belt Handbook: Tools and Methods for Process Acceleration* (2013)
- *Lean Management Systems Handbook* (2014)
- *Maximizing Value Propositions to Increase Project Success Rates* (2014)
- *Making the Case for Change: Using Effective Business Cases to Minimize Project and Innovation Failures* (2014)
- *Techniques and Sample Outputs that Drive Business Excellence* (2015)
- *Effective Portfolio Management Systems* (2015)
- *Change Management: Manage the Change or It Will Manage You* (2016)
- *The Innovation Tools Handbook, Volume 1: Organizational and Operational Tools, Methods and Methodologies that Every Innovator Must Know* (2016)
- *The Innovation Tools Handbook, Volume 2: Evolutionary and Improvement Tools that Every Innovator Must Know* (2016)
- *The Innovation Tools Handbook, Volume 3: Creative Tools, Methods, and Techniques that Every Innovator Must Know* (2016)
- *Lean TRIZ: How to Dramatically Reduce Product-Development Costs with This Innovative Problem-Solving Tool* (2017)

Introduction

This contribution to the Little Big Book series, *Project Management for Performance Improvement Teams* (PM4PITs), combines two very timely topics: Project Management and Continual Improvement, each of which has dozens of books written about it. Yet, this book is the FIRST one of which we are aware that is focused on BOTH Project Management AND Continual Improvement. It not only focuses on achieving incremental performance improvement in response to "opportunities for improvement" (OFIs) but it also delves into a more radical, disruptive, and transformative level: *innovation projects.*

PERFORMANCE IMPROVEMENT, CHANGE, AND INNOVATION

What, precisely, are *change* and *innovation* and how do they apply to *performance improvement*? We see *change* as the creative exploitation of new ideas that leads to the creation of a new or improved process, product, or service; *innovation* adds value to that change in the commercial marketplace.

Most importantly, the goal of innovation is to bring the idea to a productive position, whether this is a more profitable product (extrinsic value) or simply a better way of doing something (intrinsic value). It can improve the quality of a service, a process, or a management model. In short, it is the addition of value to stakeholders, especially the customer. If no monetary reward is achieved with the addition of customer value, we would assume that the innovation at least offers an improvement to an individual, to a group, or to society. However, we would call that "creativity," not "innovation."

Accordingly, one can categorize innovation by product innovation, service innovation, process innovation, supply chain innovation, value stream innovation, marketing innovation, or other type of innovation.

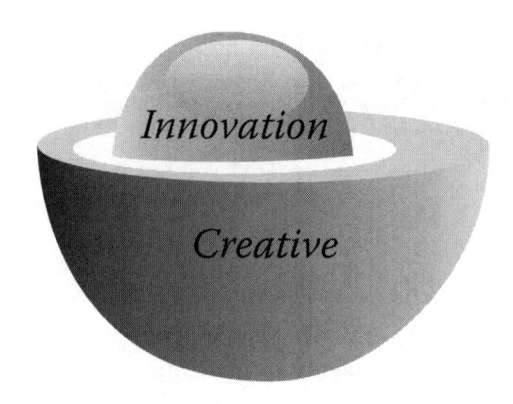

FIGURE I.1
Creativity and innovation.

Why is innovation so important? It has become a priority for all organizations—even nonprofit ones—to be or become more innovative. The majority of organizations believe that innovation is a priority and the importance of innovation is increasing, increasing significantly, and increasing rapidly. Currently, "innovativeness" has become a major factor in influencing strategic planning and it has been acknowledged that innovation leads to wealth creation in the private sector. Although efficiency is essential for business success, in the long run, it cannot, by itself, sustain organizational growth without effectiveness and strategic innovation as well.

Today, we need innovators who can help us respond to our OFIs to improve our organization's performance more than any time before; we believe that it's our Project and Program Managers who have to become those needed catalysts for change, improvement, and innovation. All organizations, in both the private and public sectors, have been feeling the impact of globalization, migration, technological and knowledge revolutions, and climate change issues, especially since the turn of the millennium. Innovation will bring added value and widen the employment base and it is imperative that the quality of outcomes in these challenging circumstances improves.

It is the central premise of this book that bridging organizational strategy with a continual stream of successful, short-term outcomes must come through the pipeline of projects and programs, each of which must be managed properly to drive incremental change and, in some cases, transformative innovation within the organization...one project at a time.

TYPES OF WORK AND WORKFORCE MANAGEMENT

Figure I.2 distinguishes four basic types of work and management in organizations (**Strategic, Operational, Project**, and. **Crisis**) and how each one treats four common variables (**Proactive & Preventive** vs. **Reactive & Corrective; Temporary & Unique** vs. **Ongoing & Repetitive; How It Treats Innovation** vs. **the Status Quo**; and, finally, Schedule Focus (**Fiscal** vs. **Short-Term** vs. **Long-Term**).

The first common variable—**Proactive & Preventive** versus **Reactive & Corrective**—shows that three of the types of work and workforce management—Strategic, Operations, and Projects—share a similar preference for a proactive and preventive approach while the fourth—a Crisis—is decidedly reactive and corrective in its initiation and execution.

Type of Work & Management	Proactive/ Preventive or Reactive/ Corrective	Temporary/ Unique or Ongoing/ Repetitive	How It Treats Innovation/ The Status Quo	Schedule Focus: Fiscal, Short-Term, or Long-Term
Strategic*	Proactive & Preventive	Ongoing & Repetitive	Welcomes It/ Visions It	Long-Term
Operational*	Proactive & Preventive	Ongoing & Repetitive	Avoids It/ Maintains it	Fiscal
*Project**	*Proactive & Preventive*	*Temporary & Unique*	*Welcomes It/ Changes It*	*Short-Term*
Crisis	Reactive & Corrective	Temporary & Unique	Avoids It/ Repairs It	Short-Term

* Can be "Corrective" during certain stages of its respective work/management.

FIGURE I.2
The four types of work and workforce management.

The second common variable—**Temporary & Unique** versus **Ongoing & Repetitive**—shows that while Strategic and Operational types of work and workforce management are both ongoing and repetitive, Projects and Crises are both typically considered temporary and unique.

The third common variable—**How It Treats Innovation** versus **How It Treats the Status Quo**—reflects the fact that Strategic and Project types of work and workforce management both welcome innovation while Operational and Crisis-oriented work and management styles usually avoid innovation since it will likely change their routines and habitual ways of performing. Yet, when it comes to the treatment of the "Status Quo" ("the way things run around here!"), each of the workforce management type differs in a fundamental way: Strategy *visions* the status quo, Operations *maintain* it, Projects *change* it, and Crises *fix* or *repair* it.

The fourth common variable—**Fiscal** versus **Short-Term** versus **Long-Term Focus**—reflects the disparity between a Long-Term (Strategic), Fiscal (Operational), and Short-Term (Projects and Crises) focus on scheduling and budgeting.

Now, having compared and contrasted these four distinct types of work and workforce management perspectives, this book focuses exclusively on the challenges and opportunities involving the third type of work and workforce management—**Projects** and **Project Management**—that are characterized by being *Proactive and Preventive, Temporary and Unique, and Change-oriented and Innovative, and by having a Short-Term schedule focus.* Finally, this book focuses on ***performance improvement*** projects rather than new product design and/or development projects.

TWO PROJECT CASE STUDY EXAMPLES

Since our goal is to provide you with practical guidance for managing your performance improvement projects to address your OFIs, we introduce you to two 21st-century projects: one for a nonprofit, healthcare program in the public sector and the other relating to a for-profit biopharmaceutical company in the private sector.

NONPROFIT/PUBLIC SECTOR CASE STUDY—HEALTHCARE DCC

Here is a summary providing the information about this organization, a federally funded grant program, and a set of strategically aligned, performance improvement projects:

- **Organization**: The Data & Coordination Center (DCC) of a Medical Monitoring & Treatment Program Consortium housed in the Preventive Medicine Department of a School of Medicine with responsibility for coordinating five geographically dispersed Clinical Centers of Excellence (CCoE).
- **Sponsor**: DCC Principal Investigator (for a NIOSH-sponsored Consortium Research Grant).
- **Primary Objective**: The Center needed to complete an extensive backlog of deliverables that had been promised in its annual progress reports and funding-renewal proposals, with some commitments pending fulfillment from the previous five years; continuation of the grant was in jeopardy. Hence, the PI created the "DCC Program Manager" position and recruited a Project Management and Lean Six Sigma expert with deep experience in facilitating cross-functional, project teams.
- **Proposed Solution**: The *"OneDCC Going Forward Plan Strategic Initiative"*—**One Team** with **One Plan** to achieve **One Goal** (e.g., demonstrate our proven scientific, data management, and coordination capabilities) driven by **One Set** of core values (**D**ata- and **D**eliverables-focused, **C**ollaboration, and **C**ommunication (DCC): completing the backlog of deliverables (referenced in the Primary Objective section) in three prioritized and time-sensitive phases containing a total of 59 projects.
- **Deliverables**: Fulfillment of:
  - The Center's nine "Original Aims"
  - The Center's six "Year #5 Aims"
  - The Center's four "Year #5 Administrative Supplement/Ex-tension Period Aims"
  - All 59 of the prioritized Projects as contained in the Center's Master Plan: phases I, II, and III.

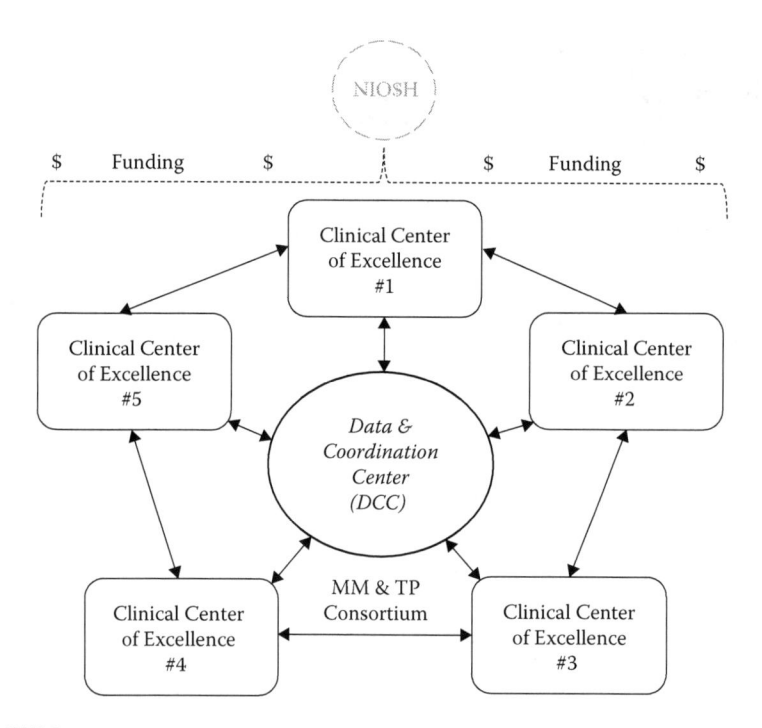

FIGURE I.3
Medical Monitoring & Treatment Program Consortium organizational chart.

Figure I.3 illustrates the Organizational Chart for this NIOSH-funded, Medical Monitoring & Treatment Program Consortium comprised of five Clinical Centers of Excellence with the Data & Coordination Center in the middle.

Figure I.4 illustrates the organizational chart for the Data & Coordination Center ONLY.

FOR-PROFIT/PRIVATE SECTOR CASE STUDY—BIOPHARMA ePROVIDE II

Here is a summary providing information about this organization and its "Top-10" Performance Improvement Project:

- **Organization**: IT Global Shared Services (GSS) Division of a Fortune 125 Biopharma Company.
- **Sponsor**: GSS Chief Information Officer (CIO).

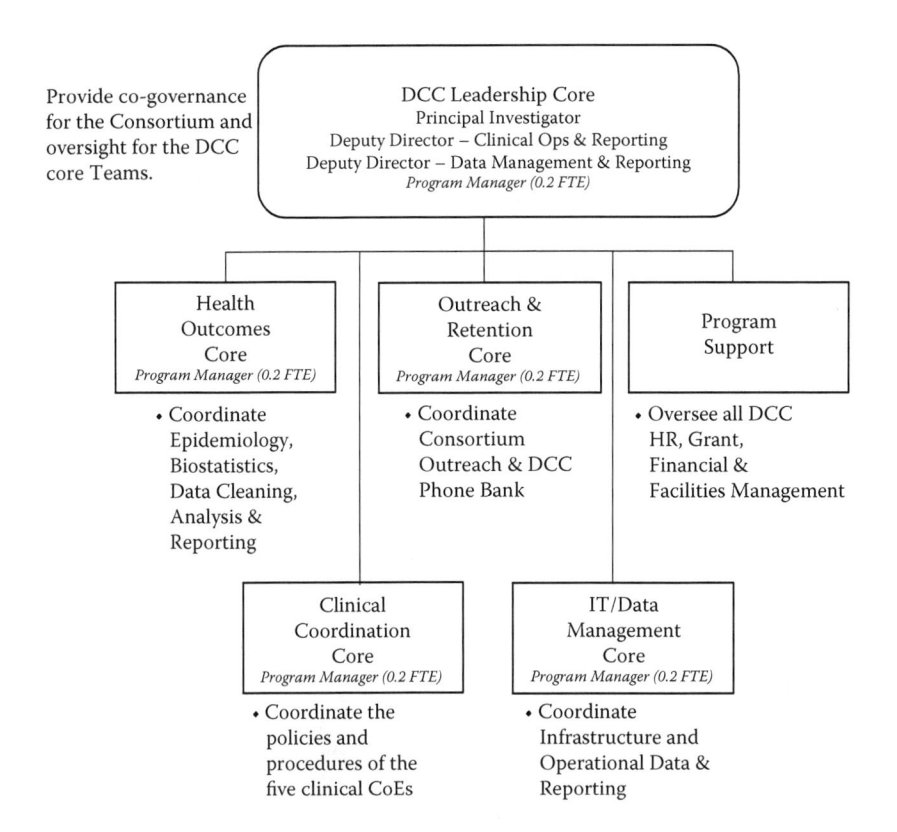

Provide co-governance for the Consortium and oversight for the DCC core Teams.

DCC Leadership Core
Principal Investigator
Deputy Director – Clinical Ops & Reporting
Deputy Director – Data Management & Reporting
Program Manager (0.2 FTE)

Health Outcomes Core
Program Manager (0.2 FTE)
• Coordinate Epidemiology, Biostatistics, Data Cleaning, Analysis & Reporting

Outreach & Retention Core
Program Manager (0.2 FTE)
• Coordinate Consortium Outreach & DCC Phone Bank

Program Support
• Oversee all DCC HR, Grant, Financial & Facilities Management

Clinical Coordination Core
Program Manager (0.2 FTE)
• Coordinate the policies and procedures of the five clinical CoEs

IT/Data Management Core
Program Manager (0.2 FTE)
• Coordinate Infrastructure and Operational Data & Reporting

FIGURE I.4
DCC organizational chart.

- **Primary Objective**: To successfully implement a "Top 10 Project" (eProvide) throughout the company within 12 months.
- **Proposed Solution**: Evaluate, select, and deploy throughout the global organization a configurable, COTS system for User Provisioning and Identity Management to replace a customized Lotus Notes system.
- **Deliverables**: Provision of:
    - A "Subject Matter Expert" (SME) to serve as the Project Manager and Coach.
    - The successful deployment of this COTS system throughout North America.
    - The completion of this "GSS Top-10 Project" on time, within budget, and with "Very Favorable" (Level 4 of 5) ratings by the stakeholders in its positive enterprise-wide impact on them.

Figure I.5 illustrates the organizational chart for this Global Shared Services division comprised of six stakeholder groups who were engaged in this "Top 10 Project."

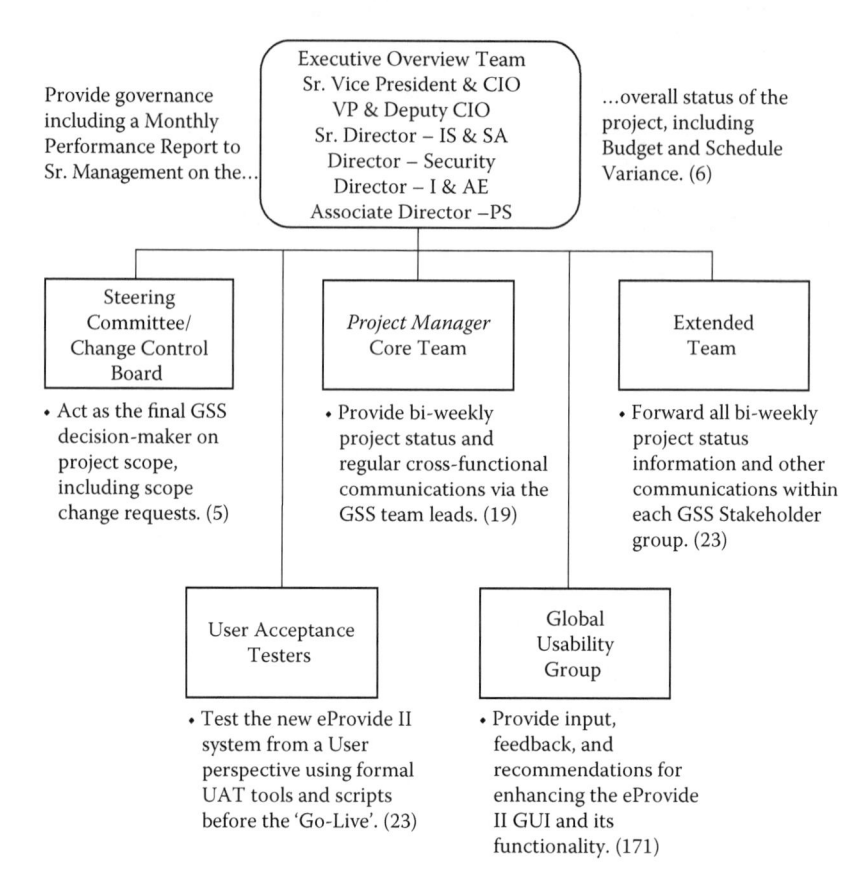

FIGURE I.5
GSS eProvide II project organizational chart.

Figure I.6 illustrates the organizational chart for the eProvide II Core Project Team ONLY.

We will revisit these two exemplary performance improvement Case Studies again at the end of Chapters 3 through 9, providing insights into how each topic was addressed and applied...both successfully (in most cases) and not so successfully (in some cases).

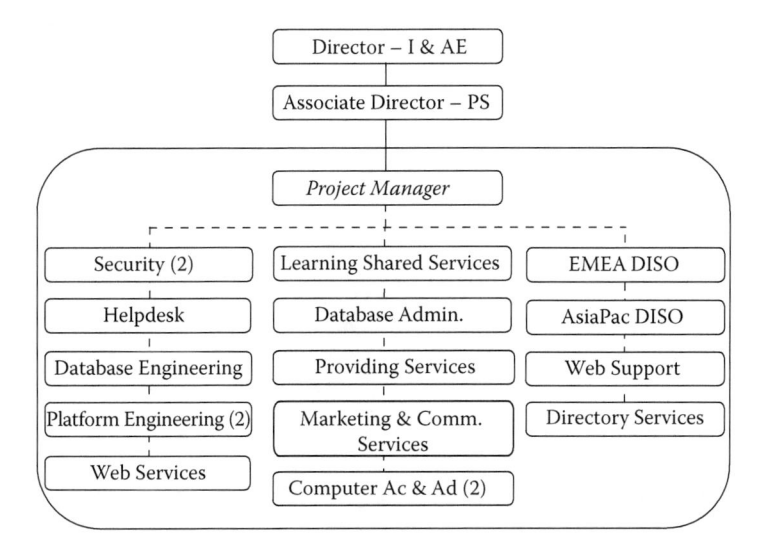

FIGURE I.6
eProvide II Core Project Team organizational chart.

1

The Traditional Frameworks for Project Management and Continual Improvement

INTRODUCTION

> "I have been assigned to manage a series of projects over the past few years, each of which has had a set of expectations—'good, fast, and cheap'—that when combined, I believe to be both unreasonable and impractical. When it first happened, I tried hard to get my project teams to achieve all three expectations, but I soon learned that there was never enough time or money to do all the work right the first time. So, we've usually resorted to doing the best we could with what we had, and failing; then, we ended up making the time and finding the money to fix it…seemingly forever! That's not what I call 'successful'!"

We've heard this "rant" over and over and, when we don't hear it, someone is usually thinking it, but doesn't want to say it out loud. Yet, a new project—especially a performance improvement project—can be the lifeblood of an organization. Without a series of successful project outcomes, such an organization lacks a secure future. Fortunately, there's a better way to manage a project in the 21st century and we'll share it with you in Chapters 2 through 11.

But first, let's make sure you're familiar with the existing, "traditional" frameworks for Project Management and Continual Improvement and, most importantly, their shortcomings.

TRADITIONAL PROJECT MANAGEMENT FRAMEWORK

Since 1996, the "traditional" or "mainstream" framework for project management, especially in North America, has been defined by the Project Management Institute (PMI)—a global, professional association for project, program, and portfolio management based in the USA—via its publication called *The Project Management Body of Knowledge* (aka the *PMBOK® Guide*).[1]

What is a project? PMI's current definition of a "project" is: "*a temporary endeavor undertaken to create a unique product, service, or result.*"[2] We believe this traditional definition falls short in one respect in that it does NOT also include the **modification** or **improvement** of an existing product, service, system, or process as the deliverable. While this traditional PM subsection does include as an example of a deliverable "an enhancement or correction to an item," it is *not* an integral part of the definition of a project, itself. Hence, you will read in Chapter 2 that our contemporary definition of a "project" is: "*a temporary endeavor undertaken to create or modify a unique product, service, system or result.*" (*Authors' note*: By "unique," we mean that there is at least ONE variable that differs from project to project (e.g., different customer, different requirements, different team members, different venue, different acceptance criteria, etc.).

Now, having clarified the definition of a "project," what is Project Management? The latest version of the traditional framework for project management defines it as "*the application of knowledge, skills, tools, and techniques to project activities to meet the project requirements*" and that it should be "*accomplished through the appropriate application and integration of the project management processes identified for the project.*"[3]

Our definition of Project Management differs slightly: "the application of knowledge, skills, tools, and techniques to balance competing constraints for a new or modified product, service, system, or result in order to meet project requirements and, when applicable, project portfolio priorities. The six primary, competing constraints are scope, quality, schedule, cost, resources, and risk." More on this in Chapter 2.

The latest version of the traditional project management framework contains three parts: a set of 13 chapters and References (Part 1); the Standard for Project Management (Part 2); and a set of six Appendices, a Glossary, and an Index (Part 3).

The first three chapters of Part 1—"Chapter 1: Introduction," "Chapter 2: The Environment in Which Projects Operate," and "Chapter 3: The Role of the Project Manager"—provide a conceptual and operational context and a theoretical foundation for projects. Then, Chapters 4 through 13 provide detailed descriptions of the following 10 Project Management Knowledge Areas: **Integration, Scope, Schedule, Cost, Quality, Resources, Communications, Risk, Procurement,** and **Stakeholder Management,** and each of its respective processes of which there are a total of 49.[4]

The "Standard for Project Management" presented in Part 2 of the current edition of the traditional framework includes an Introduction plus those same 49 processes presented in a different fashion: via their inclusion in one of five "Project Management Process Groups" (**Initiating, Planning, Executing, Monitoring/Controlling,** and **Closing**).[5]

Finally, even though they don't appear in the latest edition of the *PMBOK® Guide*, there are two other popular (but outdated) concepts that have pervaded the traditional project management training and education curricula the past 25 years. They are known as the "**Triple Constraint**" or the "**Iron Triangle.**"[6] There'll be more coverage of these two anachronistic concepts in the next section of this chapter.

What's Wrong with the Traditional Framework for Project Management?

While there's a lot to like in the latest version of the traditional framework for project management, we believe that there are at least three major shortcomings in the sixth edition.[7] As such, we believe it falls short of providing Senior Leadership, the PMO Director, and, most importantly, the Project Manager and his/her Project Team with the latest in conceptual and practical underpinnings to address the challenges and opportunities of 21st-century projects. These three shortcomings are the following:

1. The outdated "Triple Constraint" and "Iron Triangle" concepts
2. The two missing Project Management Knowledge Areas (Project Change and Project Technology Management) and their 12 combined processes
3. The ambiguous interrelationships among and interactions between the five Project Management Process Groups

Here's how we view these three shortcomings.

Shortcoming #1: The Outdated "Triple Constraint" and the "Iron Triangle" Concepts

These two concepts presume that each project has three constraints—**Scope, Time**, and **Cost**—that, together, form a three-sided object—a triangle—whose rigid shape, according to its proponents, must be maintained at all times in accordance with Stakeholder expectations or in order to deliver Quality. Hence, during the project's life cycle, if one of these constraints changes (e.g., increased Scope) so, too, must at least ONE of the other two constraints (e.g., either more Time or higher Cost) change to maintain or balance its rigid triangular shape (Figure 1.1).[8]

Unfortunately, we believe strongly that these two concepts have become anachronistic or outdated, at best, and no longer applicable, at worst. They need to be updated or replaced! We also believe that any new conceptual models should reflect the contents of the latest edition of the *PMBOK®Guide*, which actually identifies a total of SIX project constraints, NOT three, which have to be balanced on each project.[9] Yet, *none* of the hundreds of PM Training Vendors, including the PMI® Registered Education Providers (R. E. P.s), seem to have noticed and updated their curricular materials to reflect this overdue update. Hopefully, that has been officially changed with the publication of this book.

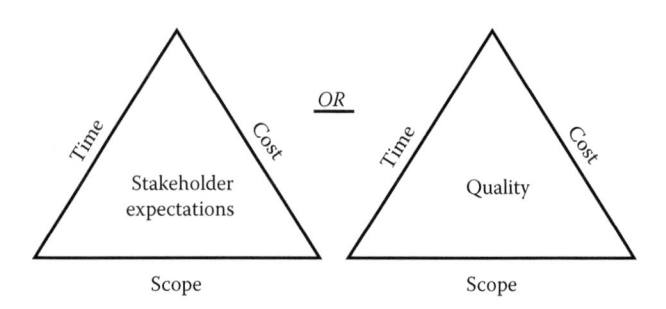

FIGURE 1.1
Triple Constraint/Iron Triangles.

Shortcoming #2: Two Missing Project Management Knowledge Areas and 12 Processes

Chapters 4 through 13 of the current edition of the *PMBOK®* *Guide* provide detailed descriptions of the 10 Project Management Knowledge Areas: **Integration**, **Scope**, **Schedule**, **Cost**, **Quality**, **Resources**, **Communications**, **Risk**, **Procurement**, and **Stakeholders** (in that order) (Figure 1.2).

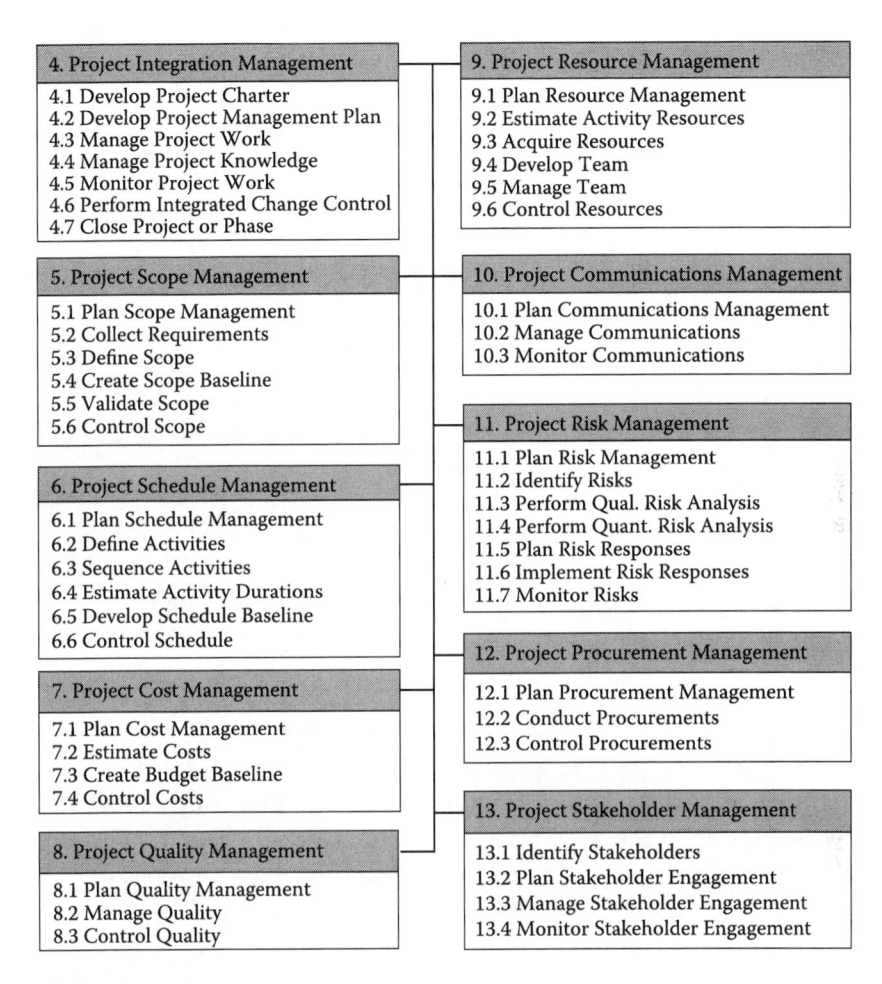

4. Project Integration Management
4.1 Develop Project Charter
4.2 Develop Project Management Plan
4.3 Manage Project Work
4.4 Manage Project Knowledge
4.5 Monitor Project Work
4.6 Perform Integrated Change Control
4.7 Close Project or Phase

5. Project Scope Management
5.1 Plan Scope Management
5.2 Collect Requirements
5.3 Define Scope
5.4 Create Scope Baseline
5.5 Validate Scope
5.6 Control Scope

6. Project Schedule Management
6.1 Plan Schedule Management
6.2 Define Activities
6.3 Sequence Activities
6.4 Estimate Activity Durations
6.5 Develop Schedule Baseline
6.6 Control Schedule

7. Project Cost Management
7.1 Plan Cost Management
7.2 Estimate Costs
7.3 Create Budget Baseline
7.4 Control Costs

8. Project Quality Management
8.1 Plan Quality Management
8.2 Manage Quality
8.3 Control Quality

9. Project Resource Management
9.1 Plan Resource Management
9.2 Estimate Activity Resources
9.3 Acquire Resources
9.4 Develop Team
9.5 Manage Team
9.6 Control Resources

10. Project Communications Management
10.1 Plan Communications Management
10.2 Manage Communications
10.3 Monitor Communications

11. Project Risk Management
11.1 Plan Risk Management
11.2 Identify Risks
11.3 Perform Qual. Risk Analysis
11.4 Perform Quant. Risk Analysis
11.5 Plan Risk Responses
11.6 Implement Risk Responses
11.7 Monitor Risks

12. Project Procurement Management
12.1 Plan Procurement Management
12.2 Conduct Procurements
12.3 Control Procurements

13. Project Stakeholder Management
13.1 Identify Stakeholders
13.2 Plan Stakeholder Engagement
13.3 Manage Stakeholder Engagement
13.4 Monitor Stakeholder Engagement

FIGURE 1.2
Ten PM Knowledge Areas.

While we are satisfied with the 10 traditional Project Management Knowledge Areas, we believe that there are still two missing—**Project Change Management** and **Project Technology Management**—each of which has six processes. We believe that these two Knowledge Areas (we call them "Performance Domains" in our contemporary framework) and 12 processes should be added to the traditional ones and we do so with our Contemporary Framework in the next chapter.

For example, when it comes to change, on the one hand, the traditional framework for project management's focus is on the impact of internal and external forces that cause change ON or TO the project. On the other hand, our contemporary framework's focus also includes the **opposite** direction as well: the impact of the project that causes change ON or TO the Requesting (Receiving) organization. Hence, without our Project Change Management Knowledge Area (Performance Domain) and its six processes, there is a sizeable gap in coverage. We include a description of them, with illustrative figures, in both Chapter 2 (Our Contemporary Framework) and Chapter 3 (Project Change Management).

We'll describe in the next Chapter what makes both of these Knowledge Areas (Performance Domains) so unique and important in our contemporary framework for Project Management as well as describe each one of them in detail in Chapters 3 and 4, respectively.

Shortcoming #3: The Ambiguous Interrelationships among and Interactions between the Five Project Management Process Groups

We believe that the interactions and interrelationships, as currently displayed in Figures 1.3 through 1.5, among and between the five Project Management Process Groups, **Initiating**, **Planning**, **Executing**, **Monitoring/ Controlling**, and **Closing**, are simply too ambiguous. Figure 1.3 illustrates the same 49 processes in the context of these five Process Groups (the columns) and the 10 Knowledge Areas (the rows).

The way the relationships between the Process Groups are illustrated is impractical for performance improvement Project Managers and their Project Teams since they are NOT fully iterative, especially between what we call the "Big 3" Process Groups, Planning, Executing, and Monitoring/Controlling,[10] in which over 93% of the processes occur!

PM Domains \ Stages	Align (2)	Plan (24)	Execute (10)	Monitor/control (12)	Confirm (1)
4—Integration	Develop Project Charter	Develop Project Mgmt Plan	Direct And Manage Project Work Manage Project Knowledge	Monitor and Control Project Work Perform Integrated Change Control	Close Project/ Phase
5—Scope		Plan Scope Management Collect Requirements Define Scope Create Scope Baseline		Validate Scope Control Scope	
6—Schedule		Plan Schedule Mgmt Define Activities Sequence Activities Estimate Activity Durations Create Schedule Baseline		Control Schedule	
7—Cost		Plan Cost Mgmt Estimate Costs Create Budget Baseline		Control Costs	
8—Quality		Plan Quality Management	Manage Quality	Perform Quality Control	
9—Resource		Plan Resource Mgmt Estimate Resources	Acquire Resources Develop Team Manage Team	Control Resources	
10— Communications		Plan Communications Management	Manage Communications	Monitor Communications	
11—Risk		Plan Risk Management Identify Risks Perform Qualitative Risk Analysis Perform Quantitative Risk Analysis Plan Risk Responses	Implement Risk Responses	Monitor Risks	
12—Procurement		Plan Procurement Management	Conduct Procurements	Control Procurements	
13—Stakeholders	Identify Stakeholders	Plan Stakeholder Engagement	Manage Stakeholder Engagement	Monitor Stakeholder Engagement	

FIGURE 1.3
The 49 processes, 10 Knowledge Areas, and five Process Groups. (Project Management Institute, *The Project Management Body of Knowledge* (aka the *PMBOK® Guide*), Sixth Edition, Newtown Square, PA, Project Management Institute, 2017, p. 25, 556. With permission.)

Notice in Figure 1.4 that the five Process Groups are depicted as a morass of overlapping and entangled lines of different types over time depicting what is called the "level of effort" for each of the five PM Process Groups: Initiating, Planning, Executing, Monitoring, and Closing within a Project or a Phase. Is the use of this Figure supposed to be a "good practice" worthy of being used as part of a project presentation to the Senior Leadership Team? We think not!

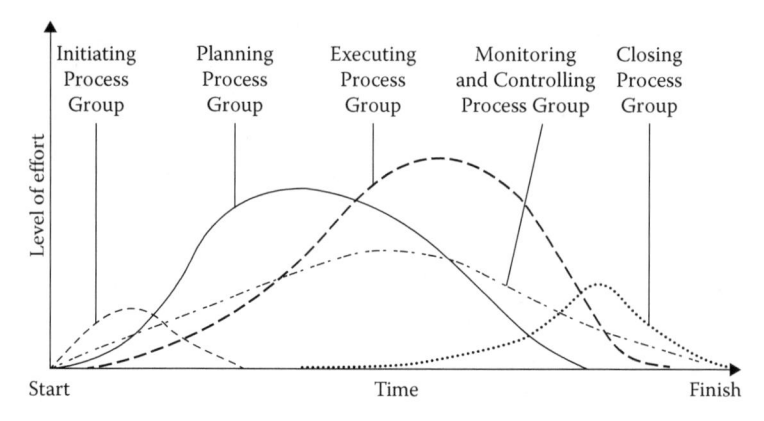

FIGURE 1.4
Overlap of the five Process Groups.

Next, notice in Figure 1.5 that the "Monitoring/Controlling" processes are depicted as an all-encompassing container surrounding the Initiating, Planning, Executing, and Closing processes, with only Planning and Executing occurring in what one could interpret as a "back-and-forth" iterative loop. In this case, it denotes that everything that happens within the project is monitored and controlled by the processes in this particular Phase/Project. This interactive ambiguity is not helpful or practical either. It also needs to be rectified. We do so in our contemporary framework for Project Management for Performance Improvement Teams summarized in Chapter 2 and explained and illustrated in Chapters 5 through 9!

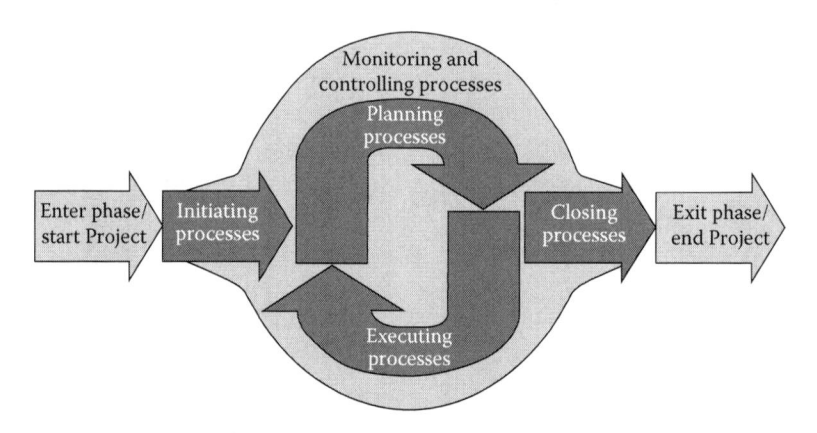

FIGURE 1.5
Interactions of the five Process Groups within each phase/project.

TRADITIONAL CONTINUAL IMPROVEMENT FRAMEWORK

The Improvement Process: How America's Leading Companies Improve Quality[11] set the foundation for continuous improvement technologies. In this book, since the two terms are often used interchangeably, we distinguish between *Continuous* and *Continual* Improvement.

Continuous Improvement is defined as a situation where

"[T]he customer's expectations are continuously getting more difficult to meet, and as a result, the process output must continue to improve." Examples would be techniques such as Total Quality Management, Area Activity Analysis, Kaizen, Activity Based Costing, Business Process Improvement, Flow Charting, etc. used as an ongoing effort to increase the efficiency of a process by eliminating waste and/or non-value added activities. It assumes a perpetual work duration."[11]

As we define **Continual Improvement**, it is focused on making improvements of various sizes achieved periodically by project teams: a traditional approach such as the "Plan–Do–Check–Act" (PDCA) cycle defined by Shewhart and the "Plan–Do–Study–Act" (PDSA) cycle defined by Deming which are used to identify and make changes that result in better, faster, less costly, or smarter project outcomes. It can be applied toward making continual, incremental enhancements in the quality of project management or the quality of a product, service, system, or result and assumes a duration that continues over a lengthy period of time but *with* intervals of interruption between projects. This book and this chapter focus on the latter term: Continual Improvement.

One of the co-authors of this book has identified over 1,400 different Performance Improvement Tools and Methodologies.[12] So, the options for identifying a Continual or Performance Improvement "framework" or "model" are nearly endless. As a result, we've narrowed it down to *one* of the most commonly used, traditional frameworks for Continual Improvement (that was originally developed to perform Problem Solving) that has been around for the past 75 years. It is known either as the "Plan–Do–Check–Act Cycle" (PDCA Cycle) or as the "Plan–Do–Study–Act Cycle" (PDSA Cycle), as well as either the "Shewhart Cycle" or "Deming Wheel," after Dr. Walter Shewhart and Dr. W. Edwards Deming, respectively, who were responsible for creating them.[13]

The PDCA Cycle is a cyclical and iterative, four-step model based on the "scientific method," originally created for problem solving or carrying out change that is often an "improvement" over something on a continual or continuing basis. While solving a problem is *NOT* continual improvement, solving a series of problems continually and using those outcomes to improve performance standards and reduce performance variance *IS* continual improvement.

One way of thinking about the differences between the two concepts is this:

On the one hand, *Problem Solving* usually creates a change based on a *negative* stimulus or *reactive* driving force (e.g., determining how to respond to a customer complaint, determining the root cause of a missed goal, performing a lessons learned after an unsuccessful bid, analyzing an inefficient process, etc.).

On the other hand, *Continual Improvement* creates a change based on a *positive* stimulus or *proactive* driving force (e.g., a higher profit margin than last year, better survey results than last quarter, a faster shipment of a package than before, a lower price than the competition…all while maintaining the same level of quality, etc.).

While Continual Improvement (CI) usually results in a change that is incremental, it could end up becoming transformational or innovative and, therefore, disruptive, too. The PDCA Cycle proceeds *clockwise*, starting with "Plan" (Figure 1.6).

Here's how each of the four steps of the PDCA Cycle is **supposed** to work:

1. *Plan*: Identify or recognize an opportunity and plan for a potential change, including the preparation of a proposal or hypothesis. (*Note*: There is a lot included in this step, which often overwhelms the user.)

aka—"Plan–Do–Study–Act"

FIGURE 1.6
The four traditional process steps (Shewhart or Deming Cycle).[9]

2. *Do*: Implement or test the proposed change or hypothesis either by carrying out a small-scale experiment or study, or by conducting a "proof of concept" pilot.

3. *Check*: Review the test/pilot, analyze the results, and use the data to analyze and identify what you've learned from it. Did you validate the hypothesis or not? Did your implemented solution actually work or not? (*Note*: This is probably the most frequently skipped step.)

4. *Act*: Take action based on what you learned in the Check phase. If the change did not work as expected, go through the cycle again with a different plan. If it was successful, incorporate what you learned from the test into other changes or on a wider scale. Use what you learned to plan new improvements or solve new problems. Then, either suspend the CI Initiative or begin the PDCA Cycle all over again by returning to "Plan." (*Note*: This step only allows for the user to Replan if the outcome of the test/pilot was different than expected.)[13]

The PDCA Cycle has been one of the key elements in most quality improvement initiatives since the early 1970s. It is, in fact, the most basic framework used to drive virtually any kind of change and is repeated iteratively for Performance Improvements until the "Solved" milestone, as illustrated in Figure 1.7.[13]

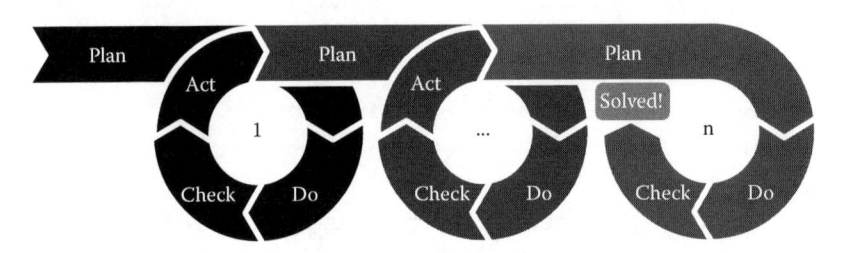

FIGURE 1.7
Iterative PDCA Cycles. (From Christoph Roser, *The Key to Lean—Plan, Do, Check, Act! April 19, 2016*. http://www.allaboutlean.com/pdca/.)

The following shows 10 detailed tasks that together expand on the activities that SHOULD take place during the traditional PDCA Cycle framework:

Phase I—Plan

- Activity 1. Identification of targets (objectives and goals)

- Activity 2. Identification of methods/procedures to achieve these targets
- Activity 3. Identify control items and methods

Phase II—Do

- Activity 4. Communicate and train associates
- Activity 5. Implement the plan (2 and 3)

Phase III—Check

- Activity 6. Check progress to plan (1, 2, 3) against targets and goals within the strategy
- Activity 7. Identify any problems

Phase IV—Act

- Activity 8. Resolve/eliminate problems
- Activity 9. Correct/modify plan (2 and 3)
- Activity 10. Standardize the improvement.

What's Wrong with the Traditional Framework for Continual Improvement?

In spite of its widespread use, much like the "traditional" Project Management Framework described above, we have encountered four serious shortcomings trying to apply this "traditional" Continual Improvement Framework that serves as the "engine" for Performance Improvement. It, too, falls short of providing Senior Management, the PMO, the Project Manager, the Kaizen Leader, the Black Belt, the Green Belt, and his/her Project Team with the latest in conceptual and theoretical underpinnings to address the unique challenges and opportunities of 21st-century projects and programs, particularly those focused on performance improvement. These four shortcomings are the following:

1. The Overloaded "Plan" Step at the frontend of the cycle
2. The Overlooked "Check" (or "Study") Step (the third step) of the cycle
3. The Inadequate "Act" Step at the backend of the cycle
4. Prevention is missing—inadequate focus on new product design and performance reliability.

Shortcoming #1: Overloaded "Plan" Step

The first step in the PDCA Cycle ("Plan") has too many issues to address and, thus, the Continual Improvement process and Performance Improvement Team tends to become overwhelmed early on. As a result, they overlook one or more of those issues (Figure 1.8).

For example, describing the purpose of "Plan" as "[r]ecognizing an opportunity and planning a change" is overly simplistic and dilutes the broader, strategic context of the undertaking and its value to the larger organization. It's become quite clear to us over the years that there's a need for another Step BEFORE this one (e.g., "Initiate," "Authorize," "Align," etc.) to address the Value Proposition, Business Case, and Project Charter associated with it as well as to identify and assign the relative Priority of the various Performance Improvement Initiatives being considered. Then, too, the Resources available may not be adequate to undertake one or more "lower priority" performance improvement initiatives.

The most important part of the improvement cycle is understanding the environment and the competition that the organization is presently facing and how it will evolve over the coming years. All too often these assessments are not conducted and when they are conducted, they are primarily directed at the executives' concerns. In most cases executives do not understand or want to understand the problems that the average line-worker is facing. This results in a list of problems and opportunities that is primarily directed at their impact upon the executive team, ignoring the employees' and customers' needs and expectations. This results in low internal morale that results in our present customers looking for another source to fill their needs and potential customers even considering doing business with another organization.

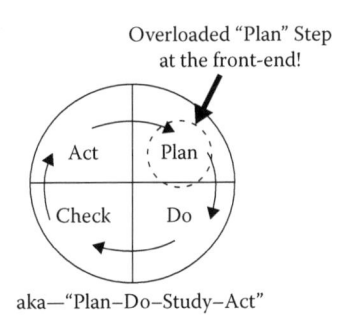

aka—"Plan–Do–Study–Act"

FIGURE 1.8
Shortcoming #1: Overloaded "Plan" Step.

"Doing a fantastic job on the wrong thing has been a tombstone for many organizations."

H. J. Harrington

Shortcoming #2: Overlooked "Check" Step

This is probably the most frequently overlooked or neglected Step by the Improvement Manager and the Continual Performance Improvement Team in the PDCA Cycle. (S)he is supposed to ask two key questions here: "Did the trial solution really work?" and "Did we prove our hypothesis?" These are two very serious questions. Based on the co-authors' collective experiences, in most cases, it does NOT work, or at least not well enough, and the Team ends up NOT proving their hypothesis (Figure 1.9).

The tendency for most performance improvement teams is to find the answer to a problem or take advantage of an opportunity, without clearly defining it first. Once a potential improvement is defined, they spend all their time and resources trying to collect data to prove that they made a good decision and that it should be implemented. For example, if the assignment is to reduce the cost of the product by 20%, which is competitive in today's marketplace, and we came up with an answer that theoretically could meet those requirements, they should set aside the tentative solution and take a fresh look at the assignment. For this new cycle, the team should establish additional complementing requirements for the end product.

For example, in this case, it could be "what needs to be done to produce the product at 20% of its present cost and decrease first-time yields by 50%?" The third one might be "How can you reduce the cycle time

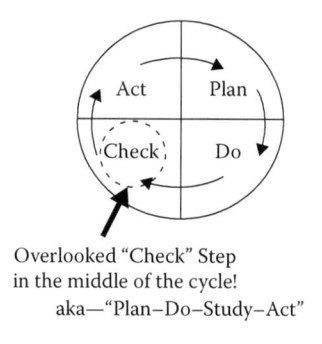

Overlooked "Check" Step
in the middle of the cycle!
aka—"Plan–Do–Study–Act"

FIGURE 1.9
Shortcoming #2: Overlooked "Check" Step.

by 40% and reduce product costs by 20%?" In each of these three cases you will get a different action plan that is specifically designed to meet established requirements. Often these three plans are combined to make a fourth plan that combines the best points in the previous three plans. Management has to learn not to be satisfied with only one answer to a problem/opportunity, but to understand what other action plans were considered.

Far too often, we have found, Senior Management is satisfied with an impressive-looking presentation, is ignorant of the unsustainable reality in the trenches, and lets the CI Project Team off the hook. They should hold them accountable for the results of this all-important "Do" step!

Additionally, they are also ignorant of the Hawthorne Effect,[14] which was first observed at the Hawthorne Works (manufacturing) of Western Electric in 1930. In many of those cases, changing something (ANYTHING) on the assembly line improved their system in the shop, regardless of what they changed—but, it turned out, *only for a short time*!

In other words, the mere attention paid during a change process where the work is being done will lead to higher productivity and better quality, for the short term, regardless of what is actually done. However, as soon as attention has moved on to something else, everything tends to revert back to the previous state. This is a very common trap. You do a project, the Key Performance Indicators (KPIs) improve, you move on, and the KPIs then revert back to what they were before. For an improvement project to work, the improvement not only has to actually work, but also has to be sustainable and continue working over an extended period of time. That is the whole idea of the "Check" part of the PDCA Cycle.

To make matters worse, the fourth (succeeding) step in the PDCA Cycle—Act—is also often ignored or neglected, leaving the organization with absolutely NO value or ROI to show for its efforts. More on that step in the next shortcoming explanation.

Shortcoming #3: Inadequate "Act" Step

Assuming we can overcome the previous two shortcomings, the "Act" step, even if it is performed, lacks the bandwidth to be able to properly create a plan or intent to confirm, standardize, or sustain the gains before closing that particular iteration of the PI initiative. Due in larger part to the overlooked "Check" step, the PI Team is unable to take action based on what it learned. As a result, we think an additional step or stage is needed at the backend of the iterative cycle as well (Figure 1.10).

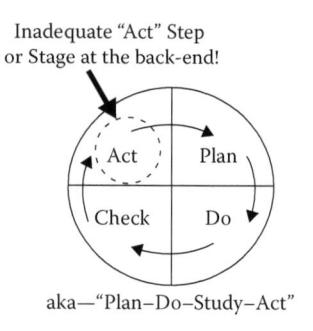

aka—"Plan–Do–Study–Act"

FIGURE 1.10
Shortcoming #3: Inadequate "Act" Step.

Shortcoming #4: Prevention Is Missing—Inadequate Focus on New Product Design and Performance Reliability

Americans are considered, by far, the world's best problem solvers. That's because they have the ***most*** problems to solve! Prevention is a technique of which we have not scratched the surface. Most quality professionals never even see the marketing specification for the product they are delivering to customers. Even ISO 9000 has removed prevention from its contents and replaced it with "risk management." (*Author's note*: As I typed risk management in this manuscript, I misspelled it as "mismanagement." Oh how true it is!)

During World War II a great deal of focus was put on developing reliable products and services and reliability engineering was an important part of every organizational structure. This emphasis lasted a few years after the war, but when products were in such demand, who had time to worry about reliability. Quality was the in-thing to do because it was readily visible before you purchase the item. Poor reliability just meant more sales. If your computer runs for three years without a motherboard or disc failure you are the exception not the rule. In either case it's cheaper to buy a new one than it is to get the old one repaired. The software companies love it because they get to sell a whole new set of software and the repairman loves it because it takes an experienced technician to install new programs and get the new computer up and running in the individual system.

A BETTER WAY

As we reassured you in the opening of this chapter that there IS a better way to manage a project—especially a Performance Improvement Project—in the 21st century and this book provides it for you.

We, the co-authors, are considered "thought leaders" by our respective peers and colleagues with over 90 years of combined leadership and experience in the fields of Project, Program, and Portfolio Management as well as in Performance Improvement, Strategic Planning/Governance, Quality, Lean, Six Sigma, Lean Six Sigma, Agile/Scrum, ISO 9001, and the Baldrige Performance Excellence Program. This experience includes serving as the International Presidents of the Project Management Institute and American Society for Quality, respectively, working in a broad variety of industries and application areas in both the private and public sectors. We've seen just about everything there is to see when it comes to these professional pursuits over the past 60 years. Yet, we see things changing in this realm and this book should help you make that transition in the remainder of the 21st century.

In the next chapter, we will introduce you to our scalable, innovative Framework merging Project Management with Continual Improvement for Performance Improvement Teams to create a contemporary model that can overcome the shortcomings we have just defined and described in this chapter.

REFERENCES

1. Project Management Institute, *The Project Management Body of Knowledge* (aka the *PMBOK® Guide*), 6th Edition, © 2017, Newtown Square, PA: Project Management Institute.
2. Ibid., p. 4.
3. Ibid., p. 10.
4. Ibid., pp. 23–25, 556.
5. Ibid., pp. 539–635.
6. WhatIs.com Definitions: http://whatis.techtarget.com/definition/triple-constraint, © 2015; Managing the Triple Constraint: https://programsuccess.wordpress.com/2011/05/02/scope-time-and-cost-managing-the-triple-constraint/ © 2011; and, Wikipedia: https://en.wikipedia.org/wiki/Project_management_triangle © 2008.
7. *PMBOK® Guide*, 6th Edition.
8. Ibid., pp. 28, 53, 542.
9. Ibid., pp. 25, 556.

10. H. James Harrington. *The Improvement Process: How America's Leading Companies Improve Quality*, p. 143, © 1987, New York: McGraw-Hill.
11. H. James Harrington, Appendix B, *Performance Acceleration Management (PAM): Rapid Improvement for Your Key Performance Drivers*, © 2013, Boca Raton, FL: CRC Press.
12. American Society for Quality, Quality Resources/Learn About Quality Portal, *Continual Improvement*. n.d., http://asq.org/learn-about-quality/project-planning-tools/overview/pdca-cycle.html
13. Christoph Roser, The Key to Lean—Plan, Do, Check, Act! April 19, 2016. http://www.allaboutlean.com/pdca/
14. Elton Mayo, *Volume 6: The Human Problems of an Industrial Civilization (The Early Sociology of Management and Organizations)*, © 1933, New York: Macmillan Company, and © 2010, New York: Routledge.

2

A Contemporary Framework for Applying Project Management and Continual Improvement for Performance Improvement Teams

INTRODUCTION

"I get frustrated trying to manage projects using the traditional 10 Management Knowledge Areas across the traditional 5 Process Groups based on the traditional 'Triple Constraint' and 'Iron Triangle' concepts. For example, what happened to Quality, Resources, and Risk as important constraints? On top of that, there are no processes for addressing the strategic impact of change and technology on my projects and my organization in the traditional frameworks. I've been looking for a truly scalable and iterative or adaptive model for ensuring that each one of my performance improvement projects is aligned with our organization's strategic priorities, especially for continual improvement initiatives, and then, getting the project work planned, executed, checked, and confirmed, iteratively, throughout the Project Management Life Cycle! Is there a better way?"

Our emphatic answer is: "YES!" There IS a better way for managing a project—especially a Performance Improvement Project undertaken to address an Opportunity For Improvement (OFI)—in the 21st century and we introduce you to this scalable and innovative framework in the ensuing pages of this book!

Over the past 30 years, we have worked on a potpourri of projects, programs, and portfolios of various sizes and in both the nonprofit and for-profit sectors in numerous *Industries* (e.g., healthcare, biotechnology, medical devices, telecommunications, construction, local/state/federal government, insurance, financial services, paper products, higher/continuing education, management consulting, and hospitality) and *Application Areas* (e.g., information technology, software development, global shared services, manufacturing, quality, research & development, new product development, marketing, sales, data management, workforce enhancement, small business development, organizational transformation, continual improvement, and turnaround management).

We've filled an assortment of positions (e.g., Project Manager, Program Manager, Portfolio Manager, Project Sponsor, CEO, COO, Director, Deputy CTO, and Senior Consultant) and leadership roles in both continuing and continual improvement settings (e.g., Total Quality Management, Business Process Improvement, Business Process Engineering, Quality Circles, Reliability Analysis and Prediction, and Knowledge Management) serving in a variety of project-related roles (e.g., Black Belt, Master Black Belt, Grand Master Black Belt, PMO Manager, Program Manager, Project Manager, Kaizen Leader, Kaizen Coach, Facilitator, Change Sponsor, and Change Agent) in more than 40 countries.

We can definitely say, "We've been there, tried that!" and we know what is likely to work and what probably won't. In this chapter, we tell you about the former: what we know is likely to work for you on most projects most of the time in today's innovation-driven work environment.

First, we invite you to review our contemporary perspective on the Project Management side.

CONTEMPORARY PROJECT MANAGEMENT FRAMEWORK

As specified in Chapter 1, as long as you can agree with us that (a) a **project** can be defined as "*a temporary endeavor undertaken to create **or modify** a unique product, service, system or result*," (b) **project management** can be defined as "*the application of knowledge, skills, tools, and techniques to balance competing constraints in order to meet project requirements and, when applicable, project portfolio priorities*," and (c) "*the six primary, competing constraints are project scope, quality, schedule, cost, resources, and risk*," we're ready to go!

Where we differ with that "traditional" project management framework, as we declared in Chapter 1, is with the following three shortcomings:

1. The outdated "Triple Constraint" and "Iron Triangle" models

2. The two missing Project Management Knowledge Areas (Project Change and Technology Management) and their 12 processes
3. The ambiguous interactions among and between the Project Management Process Groups.

CORRECTING THE SHORTCOMINGS OF THE TRADTIONAL FRAMEWORK FOR PROJECT MANAGEMENT

Correcting Shortcoming #1: The Outdated "Triple Constraint" and "Iron Triangle" Models

If you were to look carefully at a typical 21st-century project—whether it has a "Waterfall," "Iterative," "Adaptive," or "Hybrid," life cycle—you would see that there are SIX competing project constraints, *not* just three. While **Scope**, **Time**, and **Cost** are still there, so, too, are **Resources**, **Quality**, and **Risk**.

Our Question: What about these latter three Project Constraints…how are these included in the "Triple Constraint" and "Iron Triangle" models described in Chapter 1?

Our Answer: They're not! But, don't just take OUR word for this discrepancy…the sixth edition of PMI's own document (*PMBOK® Guide*) identifies all six of these competing constraints, not once but TWICE:

- "Tailoring should address the competing constraints of: **Scope, Schedule, Cost, Resources, Quality**, and **Risk**."[1]
- Managing a project typically includes but is not limited to: …balancing competing project constraints, which include but are not limited to: **Scope, Schedule, Cost, Quality, Resources**, and **Risk**…"[2]

As such, in this book, we subscribe to the notion that today's projects require project managers to address FOUR *core* constraints—Scope, Time, Cost/ Resources (combined), AND Quality—which interact together to create a fifth *residual* constraint: Risk. Thus, we have labelled our contemporary model as the "**Quadruple Constraint +1**" or the "**Flexible Quadrangle +1**" (see Figure 2.1), because it needs to be "flexible" or "negotiable" throughout the iterative Project Management Life Cycle. We'll introduce this

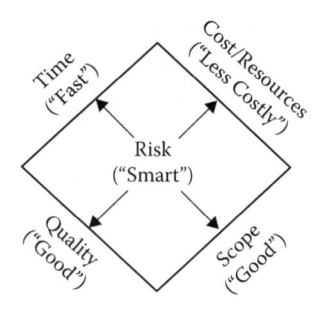

FIGURE 2.1

The Basic Quadruple Constraint +1 (aka "Basic Flexible Quadrangle +1").

interchangeable model in two Stages: first, the "**Basic**" version and, then, the "**Expanded**" version.

The Basic Version

As you can see in Figure 2.1, the "Basic Quadruple Constraint +1" is shaped like a diamond standing up on one corner with Scope, Quality, Time, and Cost/Resources around the edges and Risk squarely in the middle. Think of it as a table at which each of these constraints will be negotiated in terms of "Good" (Scope and Quality), "Fast" (Time), "Less Costly" (Cost/ Resources), and "Smart" (Risk).

Negotiated? How so? By whom? With whom? Answering these questions will require showing you our "Expanded Quadruple Constraint +1" and the Flexible Quadrangle +1 diagram.

The Expanded Version

As you can see in Figure 2.2, the "Expanded Quadruple Constraints +1" is shaped exactly like Figure 2.1 in the center, but now you see the addition of two "stakeholders": a "**Requesting Organization**" (on the left) and a "**Performing Organization**" (on the right) who are "seated" at this diamond-shaped table. One represents the fundamental interests of a "Buyer," "Customer," "End-User," or "Investor" (on the left) and the other a "Seller," "Vendor," or "Provider" (on the right) *where the Project Manager and his/her Project Team normally reside.*

Let's assume that, like any "Buyer," "Customer," "End-User," or "Investor," the "Requesting Organization" (RO) has one or more expectations about

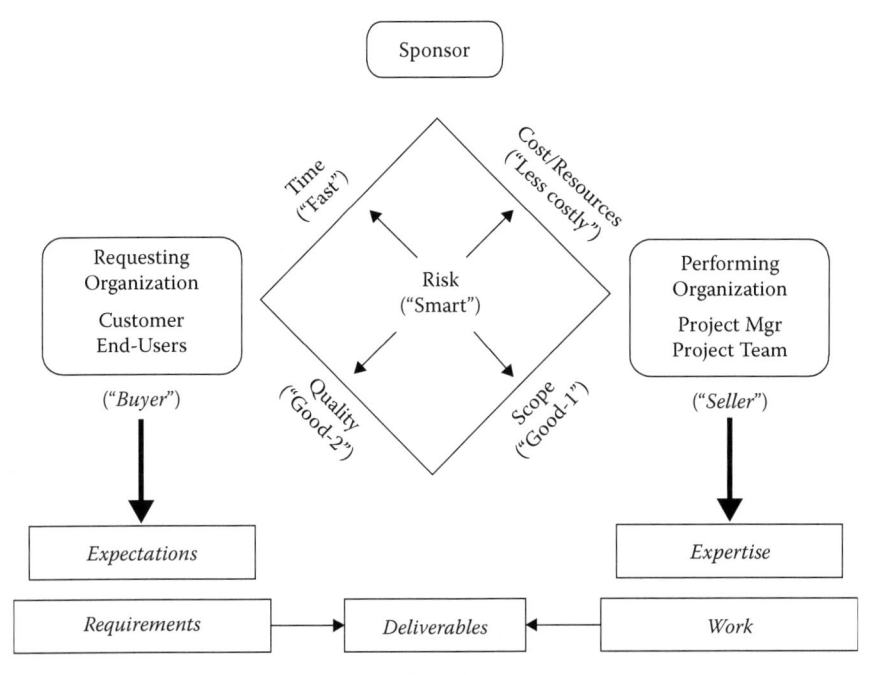

FIGURE 2.2
The Expanded Quadruple Constraint +1 (aka "Expanded Flexible Quadrangle +1").

getting a Performance Improvement Project done that will meet its value-added expectations. In addition to Quality ("Good"), the RO's expectations typically include Time ("Fast"), and Cost/Resources ("Inexpensive") but, normally, it hasn't even considered the potential Risks ("Threats") inherent in this undertaking yet.

In fact, since the RO rarely has the expertise to get the work done on its own, it must go looking for someone else who does: either inside the same enterprise or outside of it. That's where the "Performing Organization" (PO) seated on the right side of the diamond-shaped table comes in. The PO chosen *should* have the resources with the knowledge, skills, tools, and techniques (e.g., "Expertise") to submit a proposal or a quotation detailing what such an undertaking would entail, based on the RO's expectations as documented in its Requirements document.

Yet, like any "Seller, "Provider," or "Vendor," the PO has its own perspective on the very same constraints and expectations about getting that Performance Improvement Project done for the RO. But, before the PO can do so, it needs to know more about how much work is involved

(Scope). Let's "listen in" on the initial conversation between representatives for each party:

- **RO Rep (Buyer):** "Thanks for taking the time to meet with me."
- **PO Rep (Seller):** "It's my pleasure. I understand that your business unit has identified an opportunity for improvement involving one of its processes and that our Shared Services unit might have the expertise to help you fulfill your requirements in that regard."
- **RO Rep (Buyer):** "Requirements? What's a requirement?"
- **PO Rep (Seller):** "A requirement is a functional or technical characteristic of the new process you want that will, when combined with the other requirements, meet or exceed your needs and expectations. Our Business Analyst can help you identify all your initial requirements and, after you prioritize them into three levels (High, Medium, and Low) the BA will itemize and describe them in a way that could be converted by our developers into prioritized specifications. Then, we'd be able to come up with a Proposal that would include an estimate of the initial Scope, Quality, Schedule, Budget/Resource, and Risk parameters for you. At that point, we'd be able to sit down together and determine what's feasible within your constraints. We can even use an "agile approach" where you can progressively elaborate or refine your highest priority requirements or product backlog on a frequent basis over time."
- **RO Rep (Buyer):** "That'd be great! When can we start?"

(*Authors' note*: An "agile approach," used most frequently in a Software Development project environment, is described in the *Agile Practice Guide*[3] that is bundled with the sixth edition of the *PMBOK® Guide*. It is most appropriate for high-uncertainty projects that have high rates of change, complexity, and risk where it is being applied with increasing frequency.)

Hopefully, this brief conversation above will lead the two parties to a series of conversations and meetings at which a negotiated agreement (also called a "tailored outcome") will emerge.[4] For example, for any project to be successful, **both** parties must be willing to recognize that there are three possible performance improvement project outcome combinations that will be "*balanced*" or in a "*state of equilibrium.*" These three possible performance improvements are:

1. **"Good, Fast, and Smart"** (but *not* so "Inexpensive"; it'll probably cost more than desired)
2. **"Good, Inexpensive, and Smart"** (but *not* so "Fast"; it'll probably take longer to complete than desired)
3. **"Fast, Inexpensive, and Smart"** (but *not* so "Good"; it'll probably be delivered with less functionality and fewer features than desired)

    Then, there's a fourth possible project outcome combination that is "unbalanced" and that will likely be unsuccessful. This outcome combination is the one referenced back in the Introductions for Chapter 1 and this chapter. Take another look at them if you've forgotten them. Here's the toxic combination:

4. **"Good, Fast, and Inexpensive"** (but *not* so "Smart", it'll probably be imbalanced and encounter more threats and challenges than can be managed properly)

So, you may call such an undertaking a "half-baked idea," a "pipedream," "wishful thinking," "deliverable delusion disorder," or a "mission impossible." The solution may not be in line with the organization's culture and as a result, it will not be accepted by the employees. Unfortunately, these cursed project assignments have become all-too-common over the years we've been around such that one of the co-authors has a nickname for it: he simply calls it an **"Anti-Project"**!

This **"Quadruple Constraint +1"** or **"Flexible Quadrangle +1"** paradigm puts the Project Manager within the Performing Organization in a much more secure position to negotiate the necessary adjustments and make trade-offs among each project's competing objectives and alternatives to respond to BOTH types of Risk—favorable AND unfavorable:

- *Favorable*: Escalate, Exploit, Enhance, Share, and/or Accept them (*Opportunities*)
- *Unfavorable*: Escalate, Avoid, Transfer, Mitigate, and/or Accept them (*Threats*)

We believe *this* framework is the foundation upon which to manage each project in the 21st century and we aren't surprised, after we walk someone through it, when we are told that it makes so much sense. They usually want to know why someone didn't publish it sooner! On behalf of the project management professional community, we apologize for the delay, but help is here now! Please READ ON…

Correcting Shortcoming #2: The Two Missing Knowledge Areas and Their 12 Processes

As we mentioned in Chapter 1, the latest edition of the *PMBOK®
Guide* contains 10 Project Management Knowledge Areas: **Integration,
Scope, Schedule, Cost, Quality, Resources, Communications, Risk,
Procurement**, and **Stakeholders** containing a total of 49 "traditional"
processes. Yet, we think these are *inadequate* for today's innovation- and
change-driven projects. To these, we have added two more "Knowledge
Areas" (or Performance Domains, as we call them in this book) to
our contemporary Project Management Framework—**Project Change
Management** and **Project Technology Management**—each of which
has six more processes, making for a total of 12 PM Knowledge Areas
and 61 processes, as illustrated in Figure 2.3.

Project Change Management (PChM) includes the processes
required to create a Change Agenda that aligns the change objectives
and priorities of the performance improvement project with the strategic direction and expectations of the organization and its stakeholders;
identify any source(s) of resistance to that change (risk); create and
implement both an Enrollment Plan and a Change Management Plan
to address, respond to, and overcome that resistance; manage change;
check and act on the impact of any changes; and confirm the strategic
alignment of the results. PCM will be broken down and described in
Chapter 3.

Project Technology Management (PTechM) includes the processes required to create a Project Technology Management Plan that
describes and defines how the Performance Improvement Team will
plan, execute, check, and act on the implementation of the technology
(hardware and software) to be used to complete the project and to be
used in the installed design. It will be broken down and described in
Chapter 4.

Correcting Shortcoming #3: The Ambiguous Interactions among and between the Five Process Groups

Another change we have made to traditional Project Management is to
make our contemporary framework more realistic in its interactions
among and between the "Process Groups." We believe that the number of
groups, five—**Initiating, Planning, Executing, Monitoring/Controlling,**

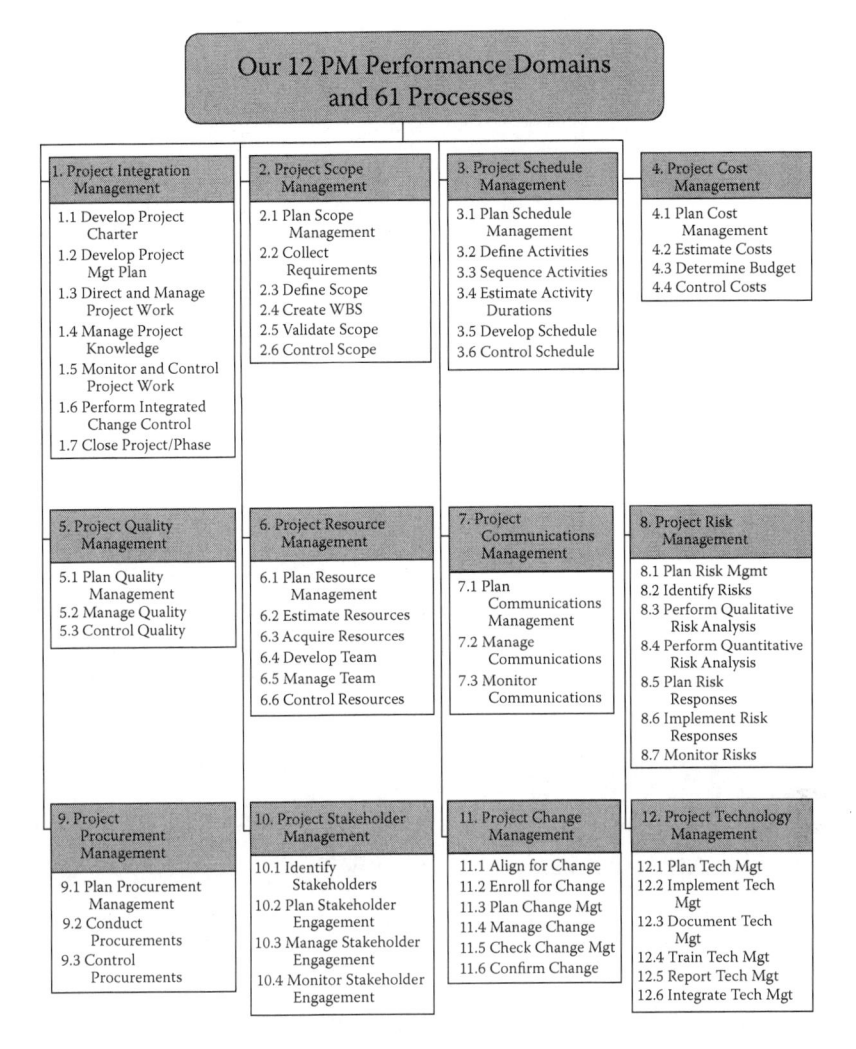

FIGURE 2.3

The 12 Project Management Performance Domains and 61 processes.

and **Closing**—is satisfactory but the way they interact should be revised (see Figures 1.3 through 1.5 to compare) to be directly iterative and adaptive in a crystal-clear model. Hence, we see them interacting as illustrated in Figure 2.4.

These new interactions among the Process Groups—especially between Planning, Executing, and Monitoring/Controlling in the center of the

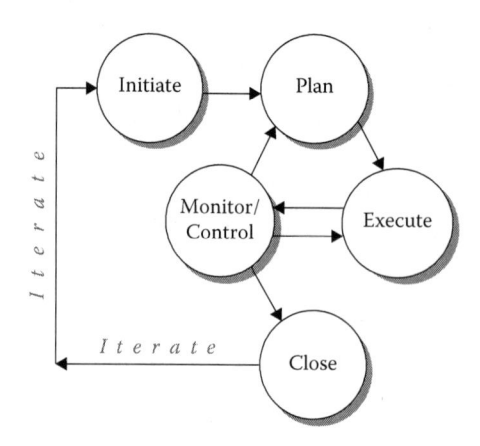

FIGURE 2.4

The five iterative PM Process Groups (arrows represent the potential iterative flows of work and data).

figure—are more consistent with the dynamic, iterative "if-then-else" interplay of 21st-century projects performed at predetermined periodic intervals (e.g., weekly, bi-weekly, monthly, etc.).

A slightly revised version of this iterative Project Management model will be merged with the Continual Improvement model to create our contemporary framework at the end of this chapter.

Speaking of continual improvement, let's now look at a refreshing, contemporary perspective on that side of our new framework.

CONTEMPORARY CONTINUAL IMPROVEMENT FRAMEWORK

Here, we shift attention to providing what we believe to be a contemporary framework for continual improvement. Where we differ with the "traditional" "Plan–Do–Check–Act Cycle" (PDCA) framework or continual improvement model is with these four shortcomings:

1. The overloaded "Plan" Step at the front-end of the cycle
2. The often-overlooked "Check" (or "Study") Step of the cycle
3. The inadequate "Act" Step at the backend of the cycle
4. Prevention—inadequate focus on new product design and performance reliability

Correcting the Shortcomings of the Traditional Framework for Continual Improvement

We believe that we can address and correct ALL four of these shortcomings by making the following three enhancements:

1. Add a new Step (Align) at the front-end *before* Plan.
2. Combine the "Check" and "Act" Steps.
3. Add a new "Confirm" Step at the backend after the "newly merged" Step.

We are leaving the "Plan" Step label but refocusing what it must achieve; we are also leaving the "Do" step and its focus as is. In so doing, we've created a five-Step "APDCC" iterative process: "Align." "Plan," "Do," "Check/Act," and "Confirm."

In addition, we believe the interactions among these five Steps should *not* be circular; they should be iterative in the same way as the contemporary project management interactions we exhibited in Figure 2.4, as illustrated in Figure 2.5.

These conditional interactions among the Steps—especially between Plan, Do, and Check/Act—in the center of the figure are more consistent with the dynamic "if-then-else" iteration of 21st-century projects.

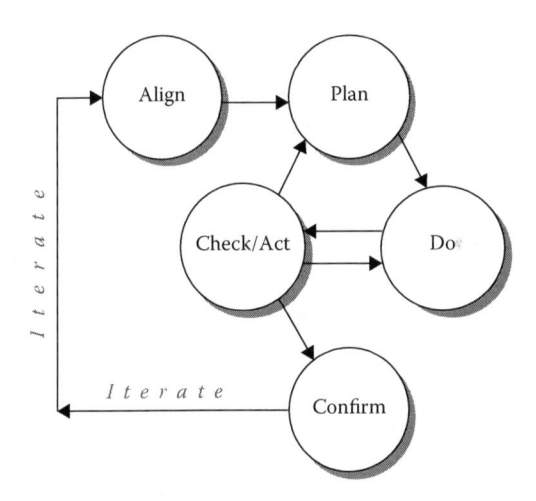

FIGURE 2.5
The new interactions of the five CI steps (arrows represent the potential iterative flows of work and data).

COMBINED CONTEMPORARY FRAMEWORK

Now, we can present our contemporary framework for combining Project Management with Continual Improvement for Performance Improvement Projects and Project Teams: the 12 Performance Domains and the five APECC Iterative Stages: "**Align**," "**Plan**," "**Execute**," "**Check/Act**," and "**Confirm**." Think of it as an enhanced "fusion" or upgrade of the two commonly used, traditional frameworks for Project Management and Continual Improvement that we presented above, illustrated in Figure 2.6.

Notice that these are referred to as Stages, rather than Steps or Process Groups, in our contemporary framework. We believe that it is better positioned to address the unique challenges of 21st-century projects, especially Performance Improvement Projects, in virtually any sector, industry or application area: healthcare, pharmaceuticals, biotechnology, financial services, information technology, manufacturing, small/medium businesses, big business, nonprofits, and government agencies to mention just a handful.

Given this contemporary framework, you can more clearly see the relationships between the 61 total processes belonging to the 12 Project Management Performance Domains and the five Stages in Figure 2.7. Applying Project Management to Performance Improvement Teams can now be done in an iterative and scalable way that matches the unique challenges and struggles facing Project Managers and their Performance Improvement Teams.

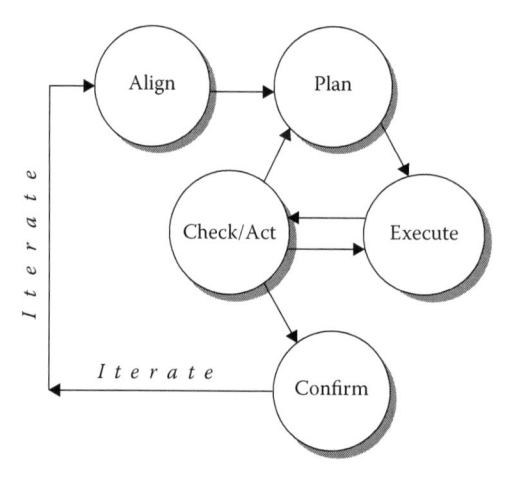

FIGURE 2.6
The APECC iterative stages of contemporary Project Management for Performance Improvement Teams (PM4PITs).

PM Domains \ Stages	Align (3)	Plan (27)	Execute (13)	Check/Act (16)	Confirm (2)
1—Integration	Develop Project Charter	Develop Project Management Plan	Manage Project Work Manage Project Knowledge	Monitor & Control Project Work Perform Integrated Change Control	Close Project/ Phase
2—Scope		Plan Scope Management Collect Requirements Define Scope Create WBS		Validate Scope Control Scope	
3—Schedule		Plan Schedule Mgmt Define Activities Sequence Activities Estimate Activity Durations Develop Schedule		Control Schedule	
4—Cost		Plan Cost Management Estimate Costs Determine Budget		Control Costs	
5—Quality		Plan Quality Management	Manage Quality	Control Quality	
6—Resources		Plan Resource Management Estimate Resources	Acquire Resources Develop Team Manage Team	Control Resources	
7—Communications		Plan Communications Management	Manage Communications	Monitor Communications	
8—Risk		Plan Risk Management Identify Risks Perform Qualitative Risk Analysis Perform Quantitative Risk Analysis Plan Risk Responses	Implement Risk Responses	Monitor Risks	
9—Procurement		Plan Procurement Management	Conduct Procurements	Control Procurements	
10—Stakeholders	Identify Stakeholders	Plan Stakeholder Engagement	Manage Stakeholder Engagement	Monitor Stakeholder Engagement	
11—Change	Align for Change	Enroll for Change Plan Change Management	Manage Change	Check Change Management	Confirm Change
12—Technology		Plan Technology Management	Implement Technology Document Technology	Train Technology Report Tech Integrate Tech	

FIGURE 2.7

The 61 processes of contemporary PM4PITs.

A Summary of Our Iterative APECC Framework

Here are brief summaries of each Stage of the Performance Improvement Project Life Cycle:

- *Align*: Ensure the strategic alignment of a new project or a new Step of an existing project by preparing a comprehensive business case that verifies the proposed project terms, conditions, and promised benefits and obtain authorization to start the project or continue to plan the next Step of the project. Identify the "opportunity for improvement," "goal," "pain point," or "problem" that is being addressed by this project and create a Change Agenda.
- *Plan*: Prepare a detailed "roadmap" for exploiting the "opportunity for improvement," "goal," "pain point," or "problem" that was identified in the Align Stage and is being addressed by this project; determine the scope, quality, schedule, budget/resource, and risk constraints for the project; refine its objectives; create an Enrollment Plan to generate support for the Change Agenda; and create a Project Management Plan supported by assorted Project Documents with the current scope, schedule, and budget performance measurement baselines to be used to guide execution.
- *Execute*: Carry out the Enrollment Plan, the Project Management Plan, and the activities defined in the Project Documents for that periodic interval (weekly, bi-weekly, monthly, etc.).
- *Check/Act On*: Review and analyze the results of Execution for the current periodic interval and identify what you've learned. Then, as per Figure 2.6, depending on the results for that periodic interval, take one of the following three actions (Act On):

  1. *If* the actual performance to date, including the value of the work performed, was consistent with the planned performance and *no* changes are needed, *then* iterate back to **Execute** to perform the work scheduled for the next periodic interval; *else*
  2. *If* the actual performance to date, including the value of the work performed, was *not* consistent with the planned performance and one or more changes are needed, *then* a Change Request needs to be created and submitted to the Change Control Board. If the Change Request is approved, iterate back to **Plan** to revise the work scheduled for the next periodic interval (and other future periodic intervals, if applicable). *Then*,

after replanning, return to **Execute** to perform the revised work scheduled for the next periodic interval; *else*

3. *If* the actual performance to date, including the value of the work performed, was consistent with the planned performance, there are *no* changes needed, and there is *no* more work scheduled for the current phase of the Project Life Cycle, or *if* the Project needs to be terminated prematurely before all of the work has been executed, *then* proceed to **Confirm** to perform its activities.

- *Confirm*: Verify that the benefits promised in the Business Case have been delivered, the Project's opportunity for improvement has been exploited, its goal has been reached, or the original problem has been solved; if the goal has NOT been achieved, recommend what Step(s) should be taken at that time. Finalize all remaining work activities to formally close the project or the current phase of the project. If there are remaining work activities for another Phase, iterate back to Align and continue.

After we cover the Project Change Management (Chapter 3) and Project Technology Management (Chapter 4) performance domains, each of these "Stages" will be defined and described in greater detail in their own respective chapter in this book, as follows:

- Align the Project: Chapter 5
- Plan the Project: Chapter 6
- Execute the Project Management Plan: Chapter 7
- Check/Act On the latest performance data: Chapter 8
- Confirm the outcome(s) (and iterate?): Chapter 9

REFERENCES

1. Project Management Institute, *The Project Management Body of Knowledge* (aka the *PMBOK® Guide*), 6th Edition, p. 28, © 2017, Newtown Square, PA: Project Management Institute.
2. Ibid., p. 542.
3. Project Management Institute and the Agile Alliance, *Agile Practice Guide*, 1st Edition, p. 2, © 2017, Newtown Square, PA: Project Management Institute and the Agile Alliance.
4. H. James Harrington and Frank Voehl, *The Innovation Tools Handbook, Volume 2: Evolutionary and Improvement Tools That Every Innovator Must Know*, © 2016, Boca Raton, FL: CRC Press.

3

Project Change Management (PCM)

INTRODUCTION

> "It's not the strongest species that survives, nor the most intelligent, but the most responsive to change."

> **Charles Darwin**

Resilient organizations have the ability to absorb high levels of disruptive change while displaying minimal dysfunctional behaviors. Resiliency requires capacity for change; it requires an organization focused on common objectives and is observed through individuals and synergistic groups having positive attitudes leading to improved results and a proactive approach to change—one that is both flexible and organized.

Many organizations "go through the motions" in predictable patterns and routines, including those that directly contribute to what made the organization great in the past. However, what made an organization great in the past may not be the case today, and may not be nearly strong enough amid increasing global competition to sustain stakeholder demands for increasingly stronger profits (private sector) or proceeds (public sector) and performance.

To break dependence on the status quo, successful Project Sponsors and Project Managers highlight the gaps between current and desired performance. These "gaps" are potential "opportunities for improvement" (OFIs). Resiliency is further enhanced by creating an appealing vision of the future state, and fostering confidence that a better future state can be achieved. Resiliency starts with a Change Agenda, an Enrollment Plan, and a Change Management Plan, while education and training enhances motivation and capacity for change.

Since change and resistance to that change occur on virtually all new initiatives, including performance improvement projects, we want to provide you with a "roadmap" for driving that change comprised of the six iterative processes of the Project Change Management (PCM) knowledge area, which we prefer to call a "Performance Domain."

As far back as the year 2000, one of the co-authors of this book co-authored another one that strongly recommended that Change Management be added as the tenth element (Knowledge Area) in the PMBOK framework.[1] Although the concept of "Change Control" and the use of a "Change Management Plan" are included in the sixth edition of the *PMBOK® Guide*,[2] Project Change Management is *not* included as a full-fledged "Knowledge Area" ("Performance Domain"). Instead, the concept of change on projects through the lens of the traditional framework is *something that impacts the project in a way that needs to be controlled*. For example, it includes:

- The premise that every project can expect to have "change requests" submitted for which there should be a "Change Requests Log" (Chapter 1).[3]
- A "Perform Integrated Change Control" process: this is a "Monitoring/Controlling" process in the Project Integration Management chapter that focuses on reviewing all change requests; approving changes and managing changes to project deliverables, organizational process assets, project documents, and the Project Management Plan; and communicating those approve/reject decisions (Chapter 4).[4]

Yet in the book from 2000 mentioned above and in this chapter, the emphasis is in the OPPOSITE direction: *how the project impacts the organization* (the Requesting or Receiving Organization). For example, we believe that the current version of the *PMBOK® Guide* doesn't go far enough in emphasizing the following:

- The premise that "projects drive change," including both "political" and "economic" changes, and that the Project Manager should consider herself/himself as a "Change Agent." (Chapter 3)[5]

As a result, we have added PCM to our contemporary model and are presenting it in this chapter (see Figure 3.1).

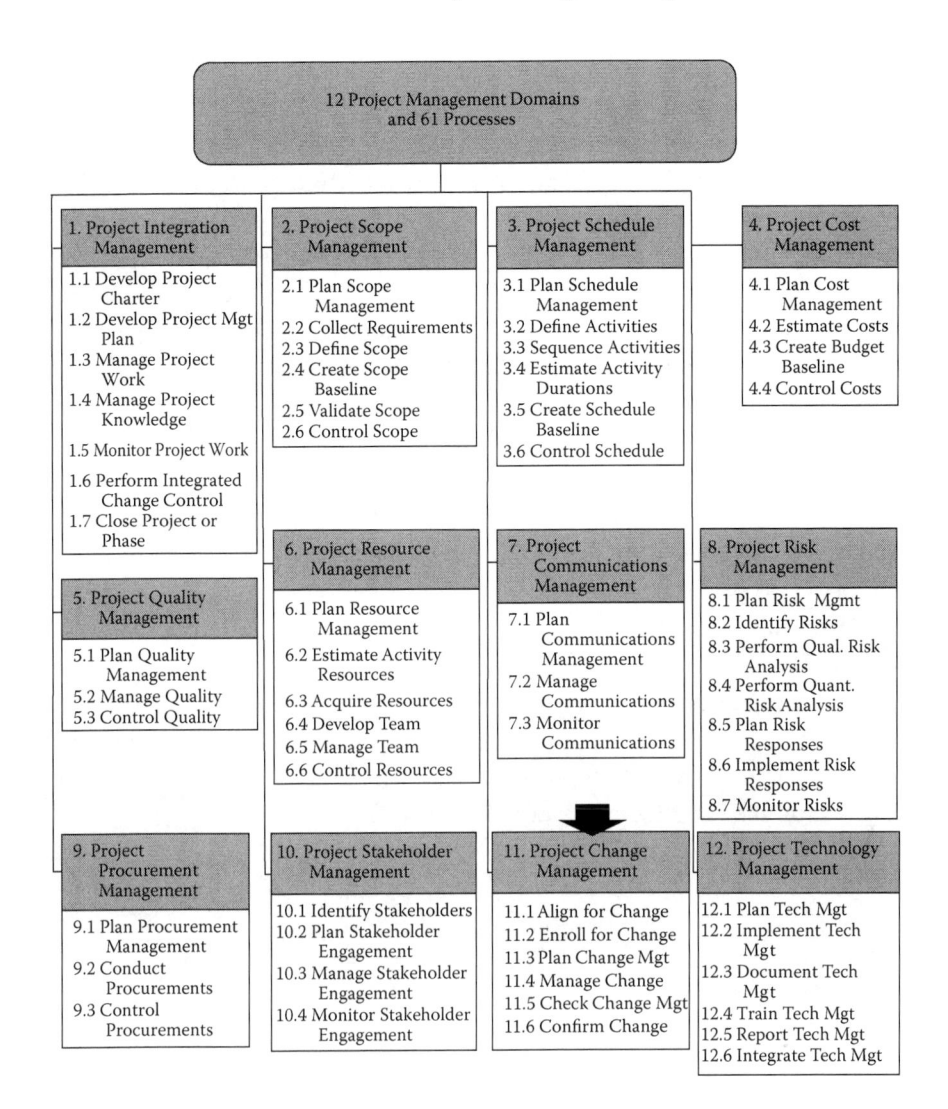

FIGURE 3.1
The 12 Project Management Performance Domains.

In this chapter, we want to help you be prepared for and overcome the resistance to change caused by your performance improvement project. It will likely surface throughout your project's life cycle as well as throughout the campaign to adopt our contemporary framework for managing performance improvement projects. We will describe each of these six PCM processes and, finally, we'll close this chapter by relating how its contents apply to one of the Case Studies presented in the *Introduction* of this book.

At first glance, PCM can appear to be a bureaucratic nightmare, but, in fact, *Gartner Group research results*[6] indicate that PCM must be an essential part of any complex improvement or change activity as a subset of Organizational Change Management. For reengineering and redesign projects, we have seen the success rate (as measured by being on schedule, within budget, and resulting in the required outcomes) jump from 60%–65% to over 90% once PCM has been added to the 10 traditional Project Management Knowledge Areas (Performance Domains).

PCM involves identifying and aligning the organization's priorities and creating a Change Agenda, an Enrollment Plan, and a Change Management Plan as part of a framework for performance management and continuous improvement. Our research shows how Project and Program Managers and their organizations can move beyond initial application of PCM into the realm of benefits realization and repeatability over time. Integrating the six PCM processes presented in this chapter on a project, especially a performance improvement project, can help you consistently realize these desired benefits.

Many project managers try to apply this 11th domain—Project Change Management—to ALL the projects on which they are working, but this could be a mistake. There are basically three types of projects where PCM should be applied. They are illustrated in Figure 3.2.

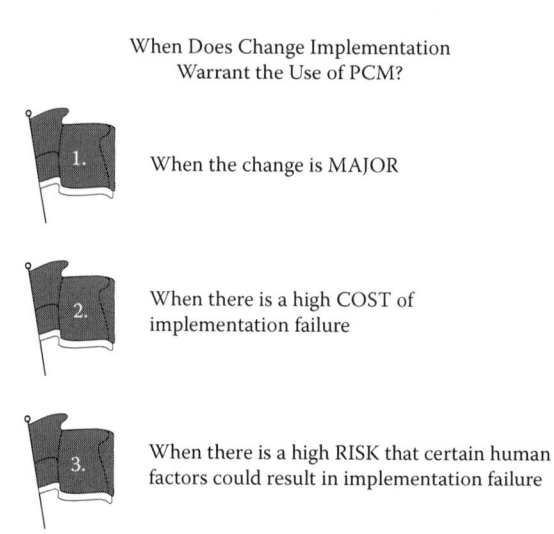

When Does Change Implementation
Warrant the Use of PCM?

1. When the change is MAJOR

2. When there is a high COST of implementation failure

3. When there is a high RISK that certain human factors could result in implementation failure

FIGURE 3.2
Three types of projects requiring PCM.

Forrester Research[7] reports that PCM approaches improve business outcomes by 10%–25% (varying by organization) on measurable factors such as:

- Process improvements and reporting automation resulting in productivity enhancements
- Improved processes for PCM resulting in improved performance against strategic organizational objectives
- Cost reductions (from implementation of Software-as-a-Service [SaaS] solutions) and avoidance (from reduced reliance on hardware and maintenance costs)
- Improved project completion success rates, from improved oversight and collaboration

Applying PCM effectively to modify the culture and prepare the impacted individuals to embrace the change results in the following:

- Better cooperation during the development of the future state.
- The approved changes are implemented faster and more successfully.
- Return on investment is much higher.
- The approved change condition lasts much longer and maintains its effective, efficient, and adaptable properties.
- Employees and management are far more satisfied with the future state.

Good practices in PCM will benefit organizations by ensuring the standardization of the following practices, typically the province of the organization's PCM, are applied to the Change Agenda:

- Providing an infrastructure for the management of projects' execution of individual Change Agendas
- Supporting review and evaluation of new initiative requests, facilitating prioritization and authorization of new projects, and allocating resources to affect change in alignment with organizational strategy and objectives
- Providing project progress reporting of critical success factor metrics, resources, expenditures, defects, and associated corrective actions to the respective Program Management Office and sustaining sponsors

- Negotiating and coordinating resources between projects in the already established processes
- Assisting with risk identification, risk analysis, and risk response planning
- Communicating risks and issues related to ongoing initiatives across projects
- Monitoring compliance of the policies to PCM and ensuring ongoing alignment with the organization's strategic objectives
- Mentoring Change Agents while developing and delivering training in process, project, and change management tools and techniques
- Providing knowledge management resources and archival services, including collection and propagation of lessons learned

PROJECT CHANGE MANAGEMENT PROCESSES

Project Change Management (PCM) includes the processes required to create a Change Agenda that aligns the change objectives and priorities of the performance improvement project with the strategic direction and expectations of the organization and its stakeholders; identify any source(s) of resistance (risk) to that change; create and implement both an Enrollment Plan and a Change Management Plan to address, respond to, and overcome that resistance; manage change; check and act on the impact of any changes; and confirm the strategic alignment of the results. (*Authors' note*: As with the other chapters, please refer to the Glossary in the back of this book for a definition of any change-related terminology with which you are unfamiliar) (Figure 3.3).

11. Project Change Management
11.1 Align for Change
11.2 Enroll for Change
11.3 Plan Change Management
11.4 Manage Change
11.5 Check Change Management
11.6 Confirm Change

FIGURE 3.3
The six processes of Project Change Management.

The *Project Change Management* Performance Domain is made up of the following six processes:

- *Process 11.1*: **Align for Change** (start at the top; identify environmental and cultural conditions that will impact the implementation and the sustained operation related to implementing the change in operations and/or personnel; identify and communicate the benefits, challenges, risks, and opportunities of the proposed change(s); and create a change agenda with a sense of urgency)
- *Process 11.2*: **Enroll for Change** (integrate the Change Agenda with the Stakeholder and Risk Management Plans; gain the support of stakeholders; provide training; create an Enrollment Plan; and, enroll key stakeholders)
- *Process 11.3*: **Plan Change Management** (conduct studies to determine what factors and individuals will have a positive and/or negative impact on the creation and implementation of proposed future state solution and create a Change Management Plan)
- *Process 11.4*: **Manage Change** (implement the Change Agenda, the Enrollment Plan, and the Change Management Plan)
- *Process 11.5*: **Check Change Management** (monitor the Change Agenda, Enrollment Plan, and Change Management Plan; monitor the status of key stakeholder support; identify any realized risks; and take the appropriate action to either replan, remanage, or confirm the change(s), as needed)
- *Process 11.6*: **Confirm Change** (confirm the strategic alignment of the change results with the Change Agenda; create a Sustain-the-Change Management Plan; need to iterate?)

These six processes occur across all five of the Stages of the iterative Performance Improvement Project Life Cycle as depicted in Figures 3.4 and 3.5.

Stage / *Domain*	Align	Plan	Execute	Check/Act	Confirm
Project Change Management	Align for Change	Enroll for Change Plan Change Management	Manage Change	Check Change Management	Confirm Change

FIGURE 3.4
The six PCM processes in the five stages of the iterative Performance Improvement Project Life Cycle.

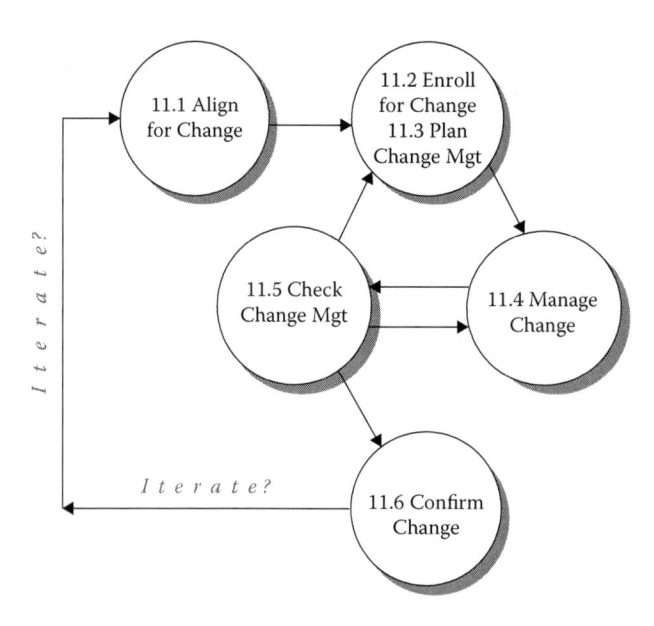

FIGURE 3.5

The six PCM processes across the five stages of the iterative Performance Improvement Project Life Cycle.

Chapters 5–9 discuss the order in which we believe these six Project Change Management processes SHOULD be performed within each Stage of the "Full" approach.

Now we present each of the six processes of Project Change Management.

PROCESS 11.1: ALIGN FOR CHANGE (start at the top; create the Change Agenda with a sense of urgency)

It's important to use a top-down approach to articulate a clear and compelling vision for where you're heading via the performance improvement project with a sense of urgency—but be realistic. Project Managers need to be aligned with and committed to that vision, understand how disruptive the change is going to be, and balance the ambition to change with the organization's capacity to absorb that change (Figure 3.6).

To attain the stated goals of performance improvement projects, individual and collective acceptance of the project's Change Agenda and expected output(s) must be obtained. Successful PCM activities looking to accomplish 95%–100% of the stated goals must first attain alignment amongst the stakeholders so that they are enrolled and all on the same page. Every sustaining sponsor is, first, a target, who is then transformed

FIGURE 3.6
Process 11.1 Align for Change.

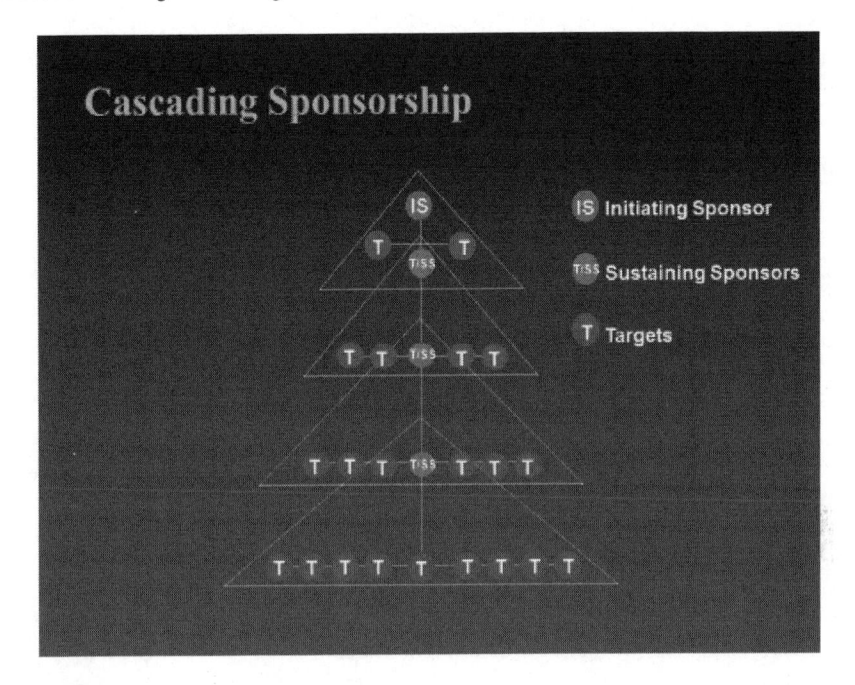

FIGURE 3.7
Cascading Sponsorship.

into a sustaining sponsor after (s)he is convinced of the value being added by the project and its deliverable(s). This transformation is called "*Cascading Sponsorship*," which is depicted in Figure 3.7.

Think of a Change Agenda as a living document—an actual working outline that summarizes and communicates the planned events and that gets disseminated (along with the Enrollment Plan and Change Management Plan, which will be described later on). Much like the purpose, agenda, and limit (PAL) guidelines used for preparing effective meeting agendas, the Change Agenda establishes the purpose, scope, and milestones of the change effort, and is distributed with the schedule of project kickoff meetings, enrollment/training sessions, and conference calls held to explain the purpose, scope, and milestones of the project and to increase the level of enrollment.

To ensure a robust alignment of the Change Agenda, begin by establishing alignment on the Leadership Team with the project goals and objectives (Cascading Sponsorship). Then, seek to enroll the Targets—the people driving and being impacted by that Change Agenda.

Once the Leadership Team gains consensus and alignment on the direction of the current project, the Change Agenda becomes a powerful force as the Project Manager or the Change Agent ensures that alignment occurs among the sustaining sponsors. When enforcing the consequences of change action, it becomes important for the Leadership Team to back one another as well as leverage interdepartmental cooperation. There is no better way to ensure adoption rates than an actively engaged and committed Sponsor placing a priority on the Change Agenda within his/her organization with a sense of urgency. With the appropriate priorities and potential positive (and negative) impacts understood, Change Agents and/ or the Project Manager will emerge as the organization takes stock of strengths, capabilities, and resources of those who can play a key role in the change effort.

The importance of starting at the top becomes evident when the Leadership Team backs up its talk with sustained positive action in promotion of the project's Change Agenda and visibly and audibly supports its goals with a sense of urgency when communicating with the staff. This is essential for Program and Project Managers when interacting with other organizational units or departments that may be contributing to or impacted by a time-sensitive project. The Leadership Team's key responsibilities are to report on the overall success of the project to stakeholders, and ensure allocation of shared resources across multiple projects and programs (new and current). At the same time, they're working to resolve multiple cross-functional issues, remove barriers, and eliminate resistance to change by promoting and enrolling others in the change effort.

One way to assess the impact of the change effort is through the organization's formal and informal social media channels, town meetings, as well as through the regular business of Project and Program review conducted by the Portfolio Management Team. The goal is to find a way for the employees and senior staff (and potentially customers) to interact with Senior Leadership, not only to demonstrate their involvement but out of a genuine active interest in the change effort borne of their own efforts and engagement.

In most groups of targets there are informal leaders or "influencers" who have an even greater impact upon their fellow employees than the management team does. It is important to identify these informal leaders or "influencers" as early as possible within the target group and involve them in the planning and implementation of the project. Once this occurs, you tend to hear statements like this: "*Tom* (the informal leader/influencer) *says this project will benefit us and not put any of us out of a job.*" If Tom says something like this, his fellow employees will likely believe it will be good for them.

PROCESS 11.2: **ENROLL FOR CHANGE** (integrate the Change Agenda with the Stakeholder and Risk Management Plans; gain the support of stakeholders; provide training; create an Enrollment Plan; and enroll key stakeholders)

Once alignment has been attained among the Leadership Team and between the organization's strategic objectives and the Change Agenda, an Enrollment Plan can be created to drive that Change Agenda and enroll the rest of the organization toward completing it (Figure 3.8).

In order to accomplish this, the Initiating Sponsor and all the Sustaining Sponsors must be enrolled first and not only commit to the goals of the project, but embrace and embody the goals of the change effort with a sense of urgency. In addition to the basic "who-what-when-where-why-and-how" of the project, the Enrollment Plan should clearly communicate the shared vision, the purpose for the change, and why it needs to be accomplished in a timely fashion. It should accomplish this by honestly outlining the opportunity (or problem), its associated challenges, and the benefits that will come about as the project goals are realized. When dealing with consequences—and they must be dealt with up front—the communication needs to strike a balance between both positive and negative consequences based on objective measures or Key Performance Indicators (KPIs).

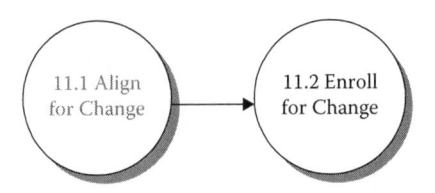

FIGURE 3.8
Process 11.2 Enroll for Change.

Each project should have an Enrollment Plan to drive alignment from the top down through the middle layers of the organization, to ensure the employees who are being impacted by the change are aware of the purposes and consequences of the change. Informed employees are empowered to help drive change or simply work collaboratively and individually within the project to best assist in the implementation and help to close the gap between mere installation (too shallow) and wider adoption (much deeper).

One practical KPI to track enrollment is through end user adoption or usage rates. For software systems, your System Administrators and DBAs should be able to provide transaction counts per day, as well as the number of unique user sessions within the system on a daily basis. The goal in calculating adoption/usage rates is to compare the number of registered users with those actually using it on a daily or weekly basis.

The Enrollment Plan is vital in setting the stage for ongoing and constant communication throughout the project life cycle, and allowing alignment to take root, which cascades into individual and collective enrollment. If lack of alignment between the business units responsible for strategy execution is one of the biggest causes of project failures, an inadequate enrollment effort makes it worse.

When you begin at the top (Align Stage) to create a Change Agenda, then proceed to create an Enrollment Plan (Plan Stage), you're much more likely to shift people's paradigms toward enrollment and engagement with a sense of urgency. It also helps to have a Leadership Team that communicates and demonstrates what it will take for everyone (themselves, included), both individually and collectively, to create dynamic vertical alignment in support of the change. With leaders at the top driving communications and benefits aligned with the change budget, the Enrollment Plan becomes more than a document of intent to explain the project to the rest of the organization. It goes beyond that by articulating that a strong alignment with the organization's strategic plan has already begun and helps the audience understand how they can get on board so that they can go out and enroll their own people, too.

When creating the Enrollment Plan based on the Change Agenda, consider these five essential factors for sustainable change among the stakeholders:

1. They must know the purpose of the project and why it is important for their team and the organization ("WIIFT = What's In It For Them").

2. They must receive communication outlining the scope and, to establish a sense of urgency, the key milestones of the project.
3. They must be enrolled in the essential elements of the project relevant to their respective roles.
4. They must understand the Critical Success Factors (CSFs) and potential areas of risk, and participate in taking reasonable actions assigned to ensure success by being prepared for the potential risks.
5. They must be assigned responsibility for the portions of the change related to them. This ensures direct accountability throughout the project life cycle.

Every person in the organization needs to have their roles clearly defined no matter how great or how small: from Project or Program Manager, facilitators, team leaders, team members, Change Agents, subject matter experts, end users, and support staff. Everyone in the affected parts of the organization must be enrolled in the change effort.

Furthermore, employees must have an incentive to change, a motivation for action, and an understanding of the risks and consequences of both failure and success. To ensure alignment with the Change Agenda, the Enrollment Plan should contain a resource or stakeholder matrix identifying the people driving the change and being impacted by it to ensure they are adequately motivated and inspired to act.

By clearly articulating the purpose of the change, its scope and consequences, and its schedule within the Enrollment Plan, the stage has been set for the organization to deal with and ultimately embrace change via the creation of a Change Management Plan.

PROCESS 11.3: PLAN CHANGE MANAGEMENT (create the Change Management Plan)

To briefly review: the Change Agenda establishes the purpose, scope, and milestones of the change effort for the initial alignment of the project; and the Enrollment Plan details how the enrollment or engagement of stakeholders will be carried out in order to facilitate the change(s) being CAUSED BY the performance improvement project and its outcome(s) on the target audience.

The purpose of those first two Project Change Management processes is to establish a high-level, two-way dialog, and to engage the Leadership Team in the concept and benefits (and challenges) while at the same time enrolling them in the effort, explaining their respective roles, and

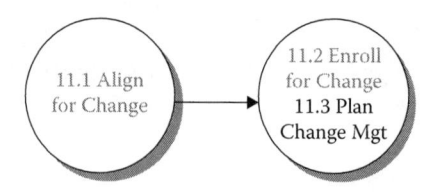

FIGURE 3.9
Process 11.3 Plan Change Management.

showing them how they will be active participants in crafting the plan for managing change. In this third process (Plan Change Management) the initial detailed Change Management Plan is prepared (Figure 3.9).

As defined earlier in this Chapter, the sole output for this process is a Change Management Plan. It is a component of the overall Project Management Plan, which defines two perspectives as to how the Project Manager will (1) Monitor and control changes TO the performance improvement project, especially its Scope, Schedule, and Budget performance measurement baselines[3] (that perspective is addressed in the Project Integration Management Performance Domain) and (2) Execute, monitor and control changes CAUSED BY the performance improvement project and any impacts on the Receiving Organization, especially the Change Targets within that organization (this perspective is addressed here).

As a result, the focus of this process is on setting the stage for executing, monitoring, and controlling changes **caused by** your project that could impact on the Receiving Organization (RO), by creating a Change Management Plan to be used in concert with your Stakeholder, Communications, and Risk Management Plans, and your Risk Register.

Change Agents are typically the ones who know best how the change may impact people and processes. The Change Management Plan (deployed in conjunction with the Enrollment Plan) should help to identify and prepare for the risks associated with the change. A balance needs to be struck between the desire for change and the organization's ability to embrace those changes. The Change Agent needs to work closely with the Project Manager to ensure that the stress related to change does not overload the targets, driving them into "future shock" from "change overload." If this occurs, the project will have difficulty meeting its commitments and the targets will become dysfunctional in relation to current activities (Figure 3.10).

By definition, each project must be schedule-bound to avoid impacting the milestones and resources of other projects that may be related to or dependent upon it. One of the goals of most projects is completion within

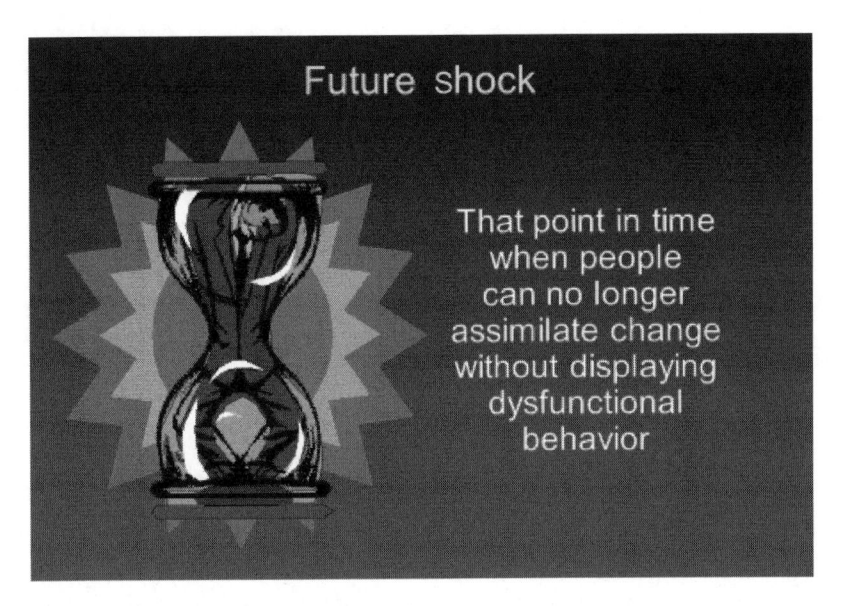

FIGURE 3.10
Future shock from change overload.

a specific timeframe, which begins the negotiation process for allocation of resources. The greatest overall risk that a Project Manager faces is *an inadequate initial alignment*, which causes the project to be misaligned throughout the project life cycle. Projects face the risk of competing for inadequate resources, budgetary constraints, and schedule constraints (missed delivery and scope creep).

From a Project Change Management perspective, doing an effective job of risk planning up front can save a tremendous amount of time down the road. This planning should result in inputs to the Plan Risk Responses process, which produces one or more of five strategies for dealing with potential negative risks (threats) (**Escalate, Avoid, Transfer, Mitigate,** or **Accept**) and five strategies for dealing with potential positive risks (opportunities) (**Escalate, Exploit, Share, Enhance,** or **Accept**).[8]

The prioritization of any change-related threats, ranging from inadequate adoption, knowledge transfer failure, all the way to employee exodus and catastrophic system failure, is critical during the Plan Stage of the iterative Project Management Life Cycle.

Risk is one of several factors the Project Team uses to evaluate the impacts of sweeping change to bring about the realization of implementing of 95%–100% of a project's stated objectives.

When analyzing potential impacts and seeking to mitigate unwarranted levels of strain within the organization, factors for consideration include urgency, risk, benefits, schedule, budget, and impact. When assessing impact, consider both impacts on the customer (beneficial) as well as on/within the organization (less desirable as this represents additional resource requirements).

Every project should have a risk analysis prepared for it—both qualitative and quantitative. There are several methods to analyze and assess risk, depending on the nature of the project and the type of data available. Following is a partial list of some of the more commonly used risk analysis and assessment methods available today:

- The Risk Probability and Impact Matrix assesses potential fail points and risks for their likelihood of occurrence and the impact they may have on the project.
- Monte Carlo simulation software applications (or spreadsheets) mathematically predict failures or overruns in complex, dynamically changing systems and environments.
- The Failure Modes Effects Analysis (FMEA) method is effective for group brainstorming on potential failures by assigning a Risk Priority Number (RPN) to potential impacts with higher likelihood and severity, compounded by lower detectability. For example, a project may have a risk associated with user adoption. With mid-level severity of five out of 10, likelihood of occurrence at five out of 10, and a 50% chance of detecting and preventing the issue, the Risk Priority Number (RPN) would be 125 (derived from SEVERITY × OCCURRENCE × DETECTION or $5 \times 5 \times 5 = 125$).
- The Harmful/Useful (HU) diagram method is used to analyze a particular product and/or solution to define harmful and useful aspects of the proposed product or solution. Each of the harmful aspects represents a risk or negative aspect to the proposed action. Analyzing these harmful aspects and implementing useful aspects to offset them is another way of identifying risks and mitigating their impact.

Whichever method is used for the change-driven risk analysis, the various risks should be prioritized, with actions identified to respond to the higher priority risks. These risk response actions should be integrated into the project's Risk Register, the overall Project Management Plan, the

Performance Measurement Baselines, and, later on, the various progress reports to assess shared impacts.

PROCESS 11.4: MANAGE CHANGE (implement the Enrollment Plan and the Change Management Plan to carry out the Change Agenda)

At this point, the Project Manager and the Team will have completed the Align and Plan Stages and have the Enrollment Plan and Change Management Plan ready for implementation. If so, they should be ready to perform the Manage Change process in the Execute Stage which is to carry out the Change Agenda, the Enrollment Plan, and the Change Management Plan for that periodic interval (weekly, bi-weekly, monthly, etc.).

It's extremely important to recognize at this crucial juncture that the Change Management Plan in conjunction with the Enrollment Plan needs to be implemented as quickly as possible. Why? Because if you wait too long, the rumor mill will already have gained inroads to the target audience, perhaps convincing them that the proposed changes are *not* in their best interests.

For example, consider Mary, who is talking with Tony while waiting in the salad buffet line in the cafeteria: "I heard my boss talking about this new performance improvement project and he was saying that it would decrease our workload."

Tony then talks to Ruth while eating lunch, "I understand that this new project is going to get rid of 15% of our direct employees."

Ruth talks to Jane, "I heard that the new project when implemented will cause anyone who has been here for less than two years to be laid off."

Jane talks to Mary, "I heard that this new project will decrease our workload so that most of us will get laid off."

Mary's answer to Jane, "You must be right. I heard my boss talking very confidentially about it to his manager. How can we keep it from happening?"

The Manage Change process implements those provisions that focus on how to execute, monitor, and control the changes CAUSED BY the performance improvement project and any impacts on the Receiving Organization, especially the Change Targets within that organization (Figure 3.11).

It should be performed in concert with these seven Execute processes described in the *PMBOK Guide*: Manage Project Work, Manage Project Knowledge, Manage Quality, Manage Resources, Manage Communications, Implement Risk Responses, and Manage Stakeholder Engagement.[9]

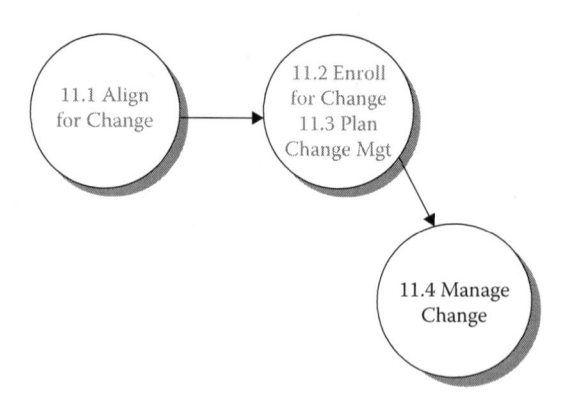

FIGURE 3.11
Process 11.4 Manage Change.

PROCESS 11.5: CHECK CHANGE MANAGEMENT (Monitor the Change Agenda, Enrollment Plan, and Change Management Plan; monitor the status of key stakeholder support; identify any realized risks; and take the appropriate action to either replan, reexecute, or confirm the change(s), as needed).

At this point, the Project Manager needs to review and analyze the results of the Manage Change process for that periodic interval (weekly, bi-weekly, monthly, etc.) and identify what (s)he has learned.

Then, depending on the results for that particular periodic interval, (s)he should take one of the following three "if-then-else" iterative actions, as per Figure 3.12.

1. *If* the change that was planned is still intact and no adjustments to the plan are needed, *then* iterate back to **Execute** to perform the work scheduled for the next periodic interval; *else...*
2. *If* the change that was planned is in jeopardy and a risk response must be implemented or one or more adjustments to the plan are needed, *then* iterate back to **Plan** to revise it for the next periodic interval. Then, proceed to **Execute** again to perform the work scheduled for the next periodic interval; *else...*
3. *If* the change that was planned has been completed and there is no more work scheduled for the current Phase or for the project, or if the project needs to be terminated prematurely due to an unexpected risk event before all of the work has been performed, *then* proceed to **Confirm Change** to perform its steps.

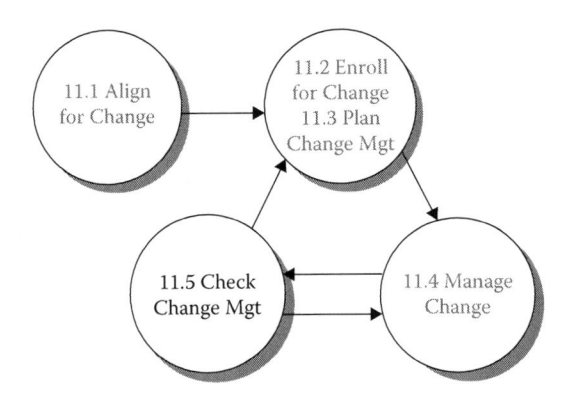

FIGURE 3.12
Process 11.5 Check Change Management.

PROCESS 11.6—CONFIRM CHANGE (confirm the strategic alignment of the change results with the Change Agenda; create a Sustain-the-Change Management Plan; need to iterate?)

Now it's time to verify or confirm that the desired change has been achieved and finalize all remaining work activities to formally close the performance improvement project or the current Phase of the project. This is where the New Process Pilot Program performance should be compared with the current process and the results made available for the executive team. If there are remaining work activities for another Phase, then iterate back to Align and resume the iterative change management model as per Figure 3.13.

Once the objectives for change have been attained and reported, it's time to close out the project. This last step is very often overlooked as a bit of change exhaustion may have set in during the course of the initiative. Understanding this dynamic will be present as your project winds down and energy naturally shifts elsewhere; plan up front on documenting the lessons learned at this Stage (transition or closure). This type of information is a critical part of the knowledge management system.

Beyond this, most of the real change takes place after the project is over. By the time the project has been implemented and the deliverable integrated within the normal day-to-day business operations, the project team has been disbanded. New (and hopefully value-adding) user habits are formed during this early post-implementation Stage. The good habits required to sustain the change(s) only become ingrained after months of continued application under normal/optimal circumstances.

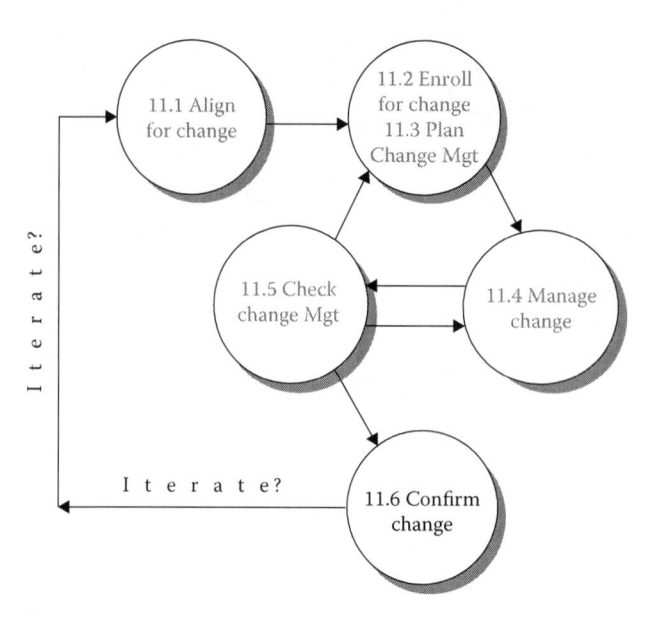

FIGURE 3.13
Process 11.6 Confirm Change (close or iterate).

Through practice and iterations of change, the return on investment will be realized once the initial successes can be replicated beyond implementation into ongoing operations and controls designed to ensure sustainability.

SUMMARY

Experts in the field of Project Change Management (PCM) universally agree that the success or failure of a change initiative like a performance improvement project is not just about aligning, planning, executing, checking on, and confirming the project that will drive the change. The Requesting (Receiving) Organization must transition to a culture that embraces that change, ensures stakeholder buy-in, and engages sponsors—both Initiating and Sustaining—to align and support the change before, during, and after its implementation. By following the iterative PCM model outlined in this chapter, organizations will be better positioned to maximize the return on investment of schedule and resources engaged for their projects and people. This results in organizational transformation by enrolling and involving impacted stakeholders in working toward solutions directly aligned with the organization's strategic objectives, one project at a time.

Applying this PCM approach directly links the Change Agenda back to the organization's strategic plan results and objectives and the Business Case upon which the project was prioritized and approved. Furthermore, PCM directly reflects the organizational strategy and objectives into its future activities. When taking a PCM-focused approach to Change Management, the organization's projects deliver a comprehensive process of reporting and assessment of value, cost/benefit, and return on investment (ROI) as per the Business Case. It also provides a broad vantage point for evaluating risk and the allocation of resources across the system.

Involving the organization's Leadership Team, Program Management Office (PMO), and impacted stakeholders throughout the five iterative Stages helps ensure project outcomes will be embraced.

Planning, measuring, and reporting on the common PCM methodology focuses on each project's Critical Success Factors (CSFs) and helps align the project with the organization's Strategic Goals and Objectives, while actively and realistically communicating the purpose, benefits, and challenges of the upcoming initiative(s).

Enrolling the impacted stakeholders includes involving them up front and providing for the mobilization and support of the organization's Information Technology/IT, Security, and Support networks. Human Resources becomes the focus mid-deployment through the development of universal training tailored to specific geographic and diverse demographic functional groups.

> "At first glance the PCM system may look like increased bureaucracy lying on top of an already slow and costly project implementation model. However, when it results in a 20% increase in the percentage of projects that are successful, and a 15% reduction in project cycle time, it really turns out to be one of the best methodologies an organization can invest in to ensure profitability and long-term sustainable results."
>
> **Dr. H. James Harrington**

REFERENCES

1. Harrington, H. James, Darryl R. Conner, and Nicholas F. Horney, *Project Change Management* (Applying Change Management to Improvement Projects), © 2000, New York: McGraw-Hill.
2. Project Management Institute, *The Project Management Body of Knowledge* (aka *PMBOK® Guide*), 6th Edition, pp. 88–89, 116, 120, 169, © 2017, Newtown Square, PA: Project Management Institute.
3. Ibid., p. 5.
4. Ibid., pp. 113–120.
5. Ibid., p. 62.
6. Voehl, Frank and H. James Harrington, *Change Management: Manage the Change or It Will Management You,* © 2016 Boca Raton, FL: CRC Press, p. 33.
7. Harrington, H. James, Conner, Daryl R., and Homey, Nick. *Project Change Management – Applying Change Management to Improve Projects,* © 2000, New York: McGraw-Hill.
8. *PMBOK® Guide*, 6th Edition, pp. 442–444, Project Management Institute, © 2017.
9. Ibid., pp. 595–611.

PROJECT CASE STUDY EXAMPLE: HEALTHCARE DCC

Here we resume describing how one of the co-authors addressed and applied Project Change Management to a set of 17 projects in a nonprofit healthcare program in the public sector that was launched in the "Healthcare DCC" Case Study at the end of this book's *Introduction*. (See the Summary with Figures I.3 and I.4.)

To refresh your memory: this was the Data & Coordination Center (DCC) of a nonprofit, grant-funded, Public Health Program that was already in serious jeopardy of losing its federal funding when one of the co-authors was recruited to assume the Program Manager role. He estimated that the employee resistance-to-change factor when he arrived was about a "9" on a scale of "1–10" and, after performing a "root-cause analysis" he found three separate but related root causes for that resistance: (1) NO Senior Leader or Leadership Team had been enforcing mutually agreed-upon, measurable Key Performance Drivers (KPDs) or Key Performance Indicators (KPIs) with individual or core accountability; (2) NO sense of urgency being communicated; and (3) a risk-averse posture by the DCC staff who were more focused on *job security* and *playing it safe* than *working together efficiently and effectively.* These three causes had been preventing them from completing their deliverables as promised to the Funding Source in the Grant Application or No-Cost Extension/Renewal Periods in a cost-effective and timely fashion. Something had to be done from a work-culture perspective!

On top of these resistance-to-change causes, after the co-author had been Program Manager for just two months, the highest ranking official—the Principal Investigator, who was an Occupational Medicine Doctor—took ill with a mysterious illness that prevented her from sitting or standing for even short periods of time. After two more months of trying to work from home lying in bed, she announced that she had been declared "permanently disabled" by her physician and would have to step down from her position indefinitely. Now, THAT adverse event was NOT in our Change Management Plan, Change Agenda, or Risk Register!

We'll now let him describe what happened in his own words:

I had been trying to meet with the four Core Leaders (see Figure I.4)— **Health Outcomes, Clinical Coordination, Outreach & Retention,** and **Data Management**—to prioritize the long-overdue deliverables from a strategic perspective, but I had been unsuccessful. Instead, I was working with the various Project Team Leads to help them put together realistic work plans and schedules for their highest-priority projects. However, it was not easy without top-down, strategic support and the needed sense of urgency to make up for lost time coming from the Principal Investigator and the Leadership Core (see the top of Figure I.4).

While the HR folks began the search for a permanent P.I. replacement, the Chair of the Department of Preventive Medicine was forced to step in to serve as the Acting P.I. It was then that I decided to shoot the moon or go for broke by implementing the six steps of Project Change Management described in this chapter. Here they are in sequential order:

1. *Align for Change* (start from the top by communicating the benefits, challenges, risks and opportunities of the proposed change to gain the support of and buy-in from the acting P.I.; create the DCC Change Agenda)

I approached the Acting P.I./Department Chair (himself an MD with a long and impressive track record) and asked him if he was willing to support my efforts to improve the work performance of the DCC and its Project Teams by helping them clear their backlog of unfulfilled deliverables in a time-sensitive fashion. He asked me what I had in mind and I shared it with him: what I called the *"OneDCC Going Forward Plan Strategic Initiative."* With the federal funding source applying its own pressure on the School of Medicine separately, it made it much easier for me to "sell" it to him. He agreed, in principle, but wanted to know more details, including the perspectives of his four Core Leaders, before making a final commitment to proceed.

2. *Enroll for Change* (enroll the DCC Core Directors; integrate the Change Agenda with the Stakeholder and Risk Management Plans; gain the support of other key DCC Staff; and create the Enrollment Plan to integrate with the DCC Change Agenda)

I made sure there was an item on the next Weekly DCC Core Leadership Team Meeting where I could share my ideas with the four Core Leaders and convince them to join the Acting P.I. (their boss) and me in announcing this strategic initiative at an upcoming Weekly DCC Staff Meeting. Given the fact that the Acting P.I. had said he would support it, they all agreed to join in but wanted to see some type of work plan with a Milestone Schedule for the various projects in their order of priority (highest to lowest). I agreed to work with them to produce that Master Plan by the next Weekly Core Leadership Team Meeting.

3. *Plan Change Management* (create the DCC Project and Change Management Plans)

So, I collaborated with each of the Core Leaders to agree upon a consensus list of priority projects and invited the Acting P.I. to the next Weekly Core Leadership Meeting to review it with us. I presented a more detailed version of the *"OneDCC Going Forward Plan Strategic Initiative"*—One Team with One Plan to achieve One Goal (e.g., to demonstrate our proven scientific, data management, and coordination capabilities) driven by One Set of core values (Data- and Deliverables-focused, Collaboration, and Communication) (DCC). (See Figure 3.14.)

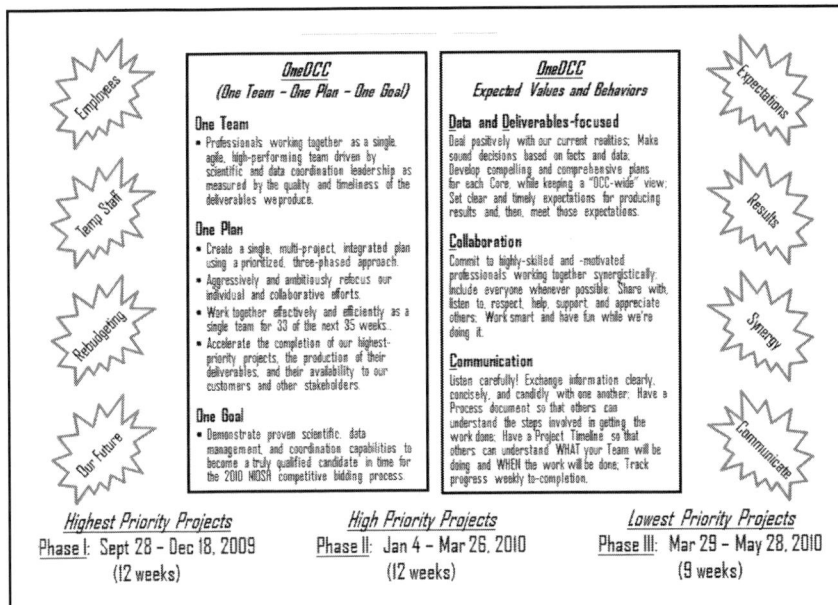

FIGURE 3.14
OneDCC Going Forward Master Plan poster.

That One Goal was to complete the backlog of deliverables in three prioritized and time-sensitive Phases containing a total of 59 projects…one project at time…over a period of 18 months (see Figure 3.15). I presented this time-sensitive Master Plan to the Acting P.I. and the Core Leadership Team at their next meeting and, after a few fine-tuning adjustments and refinements in the grouping of projects, much to my surprise, THEY APPROVED IT!

At the next Weekly DCC Staff Meeting, attended by the Acting P.I., all four of the Core Leaders, and the Program Support Leader, I presented the *"OneDCC Going Forward Plan Strategic Initiative"* to the entire DCC Staff, answered their questions, and got them to enroll or buy into it. After quite a bit of hand-wringing, hemming, and hawing, each one of those staff members present either verbally agreed or quietly consented. We got the go ahead to commence the initiative with the sense of urgency I had hoped for, which would be monitored by the Leadership Core on a weekly basis going forward.

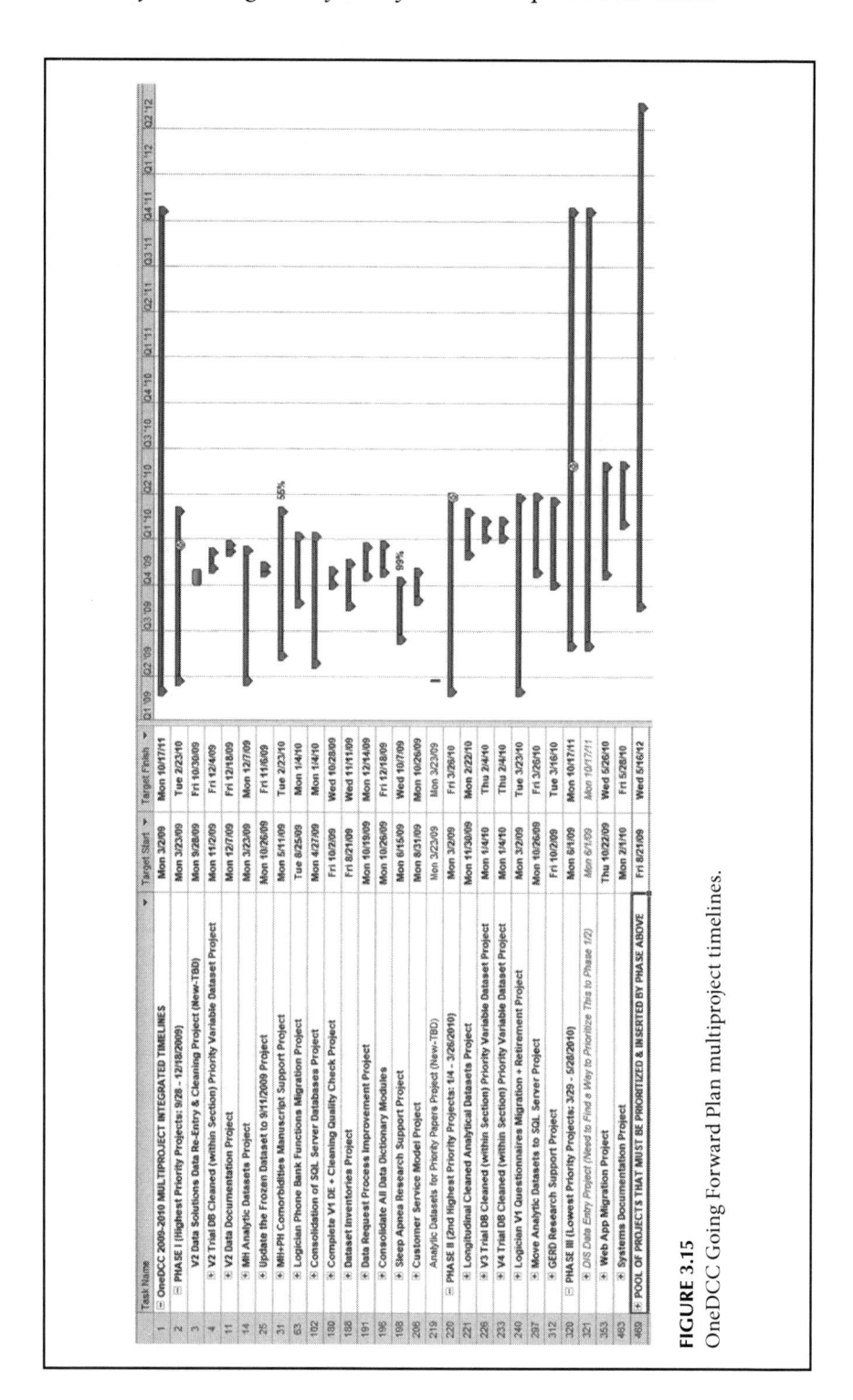

FIGURE 3.15
OneDCC Going Forward Plan multiproject timelines.

4. *Manage Change* (implement the Change Agenda, the Enrollment Plan, and the Change Management Plan)

For the next 12 weeks (September 28–December 18), we executed 17 of the highest-priority or most time-sensitive projects identified in Phase I of the *"OneDCC Going Forward Plan Strategic Initiative*," with a two-week holiday break afterwards so that we could take a "breather" before commencing Phase II following the New Year's celebration. (See Figure 3.16 for the Phase I multiproject integrated rolled-up timelines.)

5. *Check Change Management* (Monitor the Change Agenda, Enrollment Plan, and Change Management Plan; Monitor the status of key stakeholder support; identify any realized risks; and take the appropriate action to either replan, reperform, or confirm change(s), as needed)

We met each week for the next 12 weeks and, during each Weekly Meeting, we used a color-coded Tracking Gantt Chart to (1) check the progress with the Project Manager of those projects which were supposed to be "Started," "In-Progress," or "Completed," that *same* week; and (2) check the latest status with the Project Manager of those projects that were supposed to be "Started" or "Completed" the *following* week (see Figure 3.17 for sample Tracking Gantt Chart – with colors removed – as of the end of the third week of November of Year 5).

The "color-coded" set of progress/status symbols was based on the classic "Traffic Signal" approach: "Green" = on or ahead of schedule; "Yellow" = behind schedule; and "Red" = missed completion date. Since we wanted to be sure we were NOT "managing these projects by looking at the review mirror" as is common with the "traffic signal approach," we kept the focus on each project's schedule performance as trends over time, not just a single snapshot. Since everyone had had a say in finalizing these project schedules, the Funding Source was checking our progress regularly, and there was a new sense of urgency; it was going more smoothly than I had anticipated.

Believe it or not, this simple approach was sufficient to get the attention of each of the DCC Project Managers and Team Leaders. The preexisting "risk-averse," "playing-it-safe,"

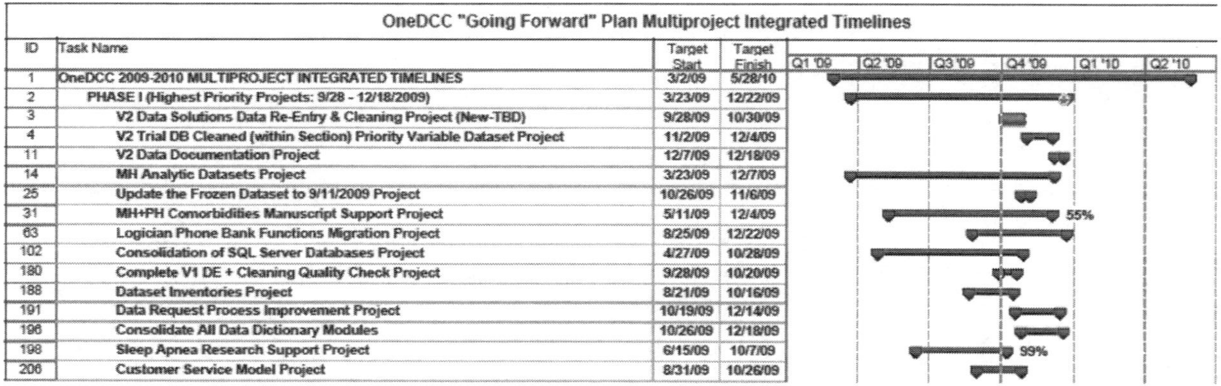

FIGURE 3.16
OneDCC Going Forward Plan Phase I rolled-up timelines.

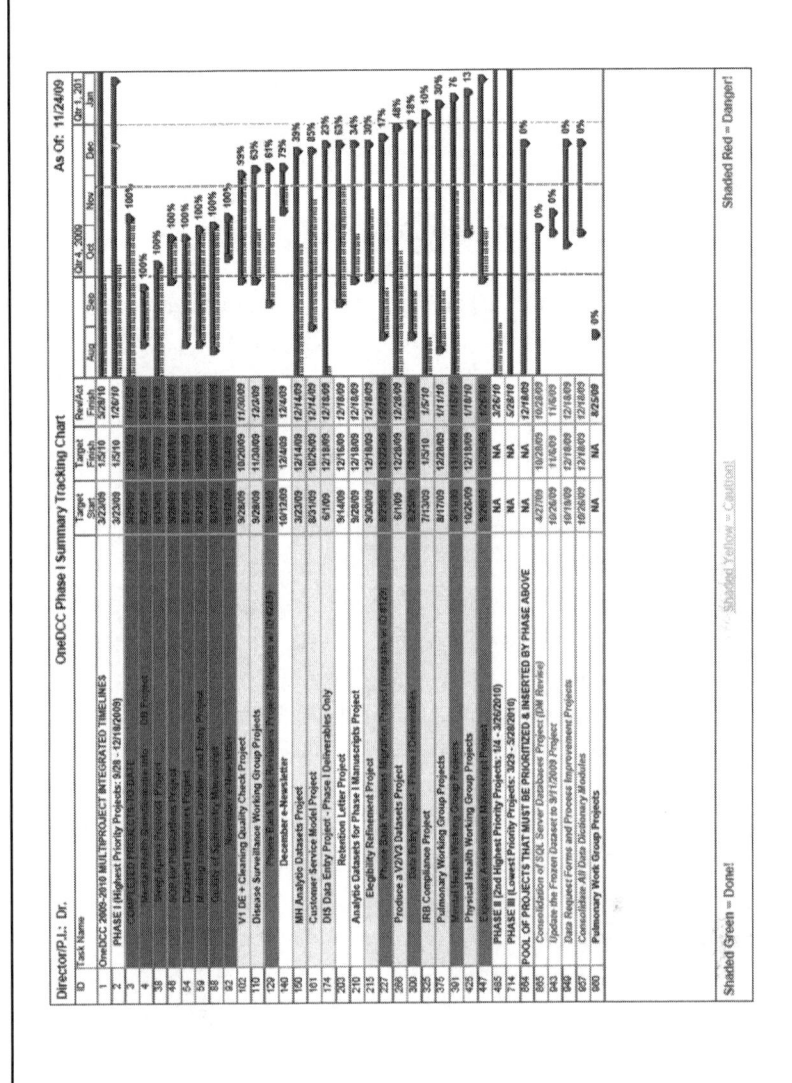

FIGURE 3.17

OneDCC Going Forward Plan Phase I rolled-up check/act.

"job-security-above-all" mentality mentioned in an early part of this Case Study actually worked in our favor at that point. Why? Because no one wanted to "suffer the embarrassment" of having their project (and, therefore, themselves) highlighted in either "Yellow" or "Red" in front of the Acting P.I. and the Core Leaders at one of the Weekly Meetings, without a legitimate reason or explanation. That might jeopardize their future "job security"! However, in those cases where there was an unforeseen schedule delay, we were able to persuade everyone to address it as OUR problem for which WE should help the Project Manager and his/her Team find and implement a solution together. It seemed to work!

While we did NOT Check on each project's budget performance during these Phase I Weekly Meetings, we DID check and act on their Scope, Schedule, and Quality performances directly, with an emphasis on Schedule due to the direct impact of project completions on our continued funding. Above all and, in the long run, our continued funding would be the ONLY budget or financial performance that meant anything to the OneDCC Team!

6. *Confirm Change Outcomes* (confirm the strategic alignment of the change outcome; create a Sustain-the-Change Results Plan to standardize the change; need to iterate?)

Much to our surprise (but pleasantly so), we successfully completed 12 of the 17 highest-priority or most time-sensitive projects for Phase I before going on that pre-scheduled two-week holiday break. The other five had slipped into January 2010. But that FAR exceeded our expectations!

So, in summary: we had created a Change Agenda with an Enrollment Plan, a Change Management Plan, a Risk Register, and a Project Management Plan for each of the 17 projects for Phase I that communicated the right level of urgency without causing panic among the DCC Staff.

We finished the Calendar Year on an upbeat note and it appeared that we had turned things around in less than a year, even with the unanticipated loss of our Senior Leader, the Principal Investigator, only six months earlier.

"A critical aspect of effective process improvement programs is a disciplined application of a rigid change management methodology."

H. James Harrington

4

Project Technology Management (PTechM)

INTRODUCTION

On the one hand, the traditional framework for project management considers project technology—both hardware/infrastructure and software—to be one of several "Enterprise Environmental Factors" (EEFs) *internal* to the organization that are conditions, *outside* the control of the project team, that influence, constrain, or direct the project.[1]

Yet, on the other hand, we know that among the many reasons so many of our projects and programs have failed to deliver the desired results and benefits is the *poor use of technology*.[2] Since project technology has become so ubiquitous in the 21st century, strategic initiatives, including performance improvement projects, shouldn't someone provide a roadmap for mastering or harnessing project technology in more detail in the form of a Performance Domain: ***Project Technology Management*** (PTechM)? Our answer is an emphatic "Yes" and we do so here!

In this chapter, we want to help you overcome the disruptive impact that technology will likely have on your project at certain Stages of its life cycle, especially those projects carried out using "virtual" or geographically dispersed Project Teams. This should be done via the creation and execution of a Technology Management Plan that will become a component of your Project Management Plan.

Organizations that have a PMO with a portfolio of projects should already have a set of project technologies in the form of a "standard" set of applications or a single system for managing projects. This will reduce the amount of time the Manager of a single project will need to devote to this Performance Domain and its six processes. We close this chapter by relating how its contents apply to one of the Case Studies presented in the *Introduction* of this book

PTechM includes the processes required to create a Project Technology Management Plan, which describes and defines how the Performance Improvement Team will plan, execute, check, and act on the implementation of the technology (hardware/infrastructure and software) to be used to complete the project.

Figure 4.1 shows the project management performance domains with Project Technology Management highlighted as the twelfth performance

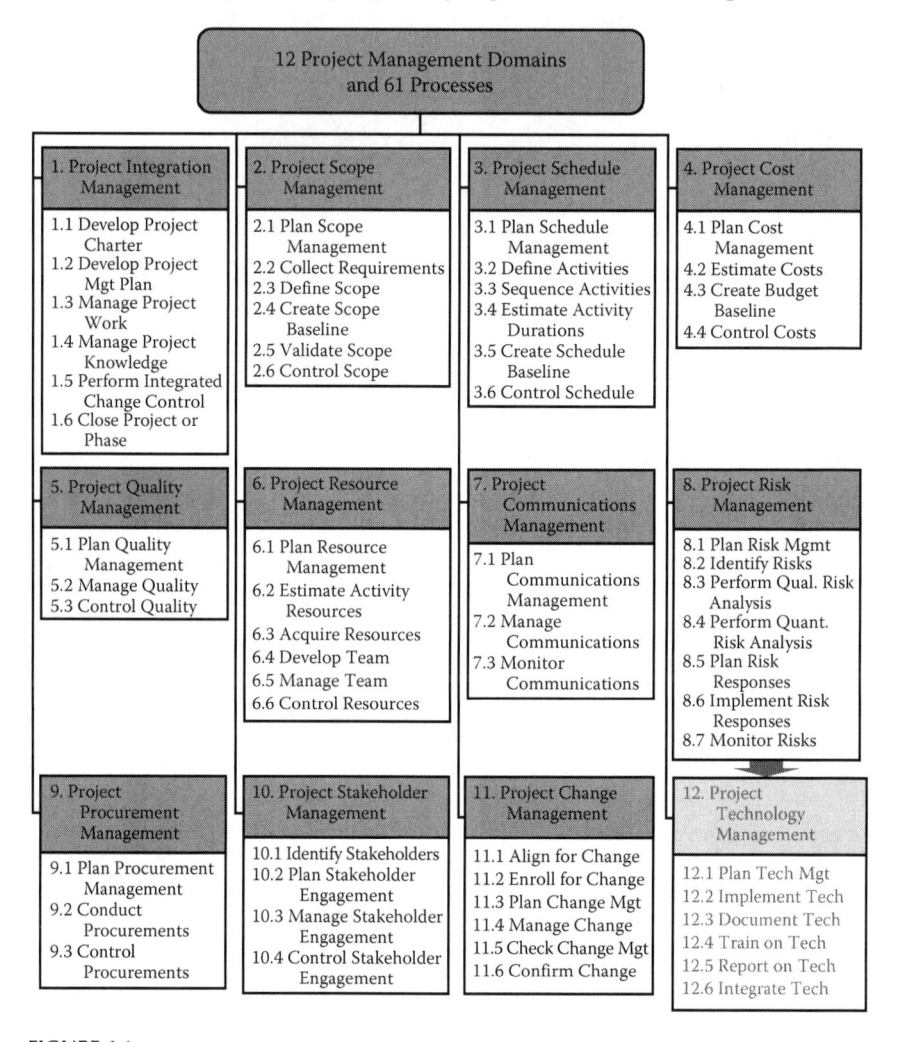

FIGURE 4.1

Project Technology Management and the rest of the 12 Project Management Performance Domains.

domain. We believe that PTechM plays a very important role in project management, continual improvement, and innovation and, as a result, it should be treated as a separate performance domain.

PROJECT TECHNOLOGY MANAGEMENT PROCESSES

As mentioned above, Project Technology Management includes the processes required to create a Project Technology Management Plan, which describes and defines how the Performance Improvement Team will plan, execute, check, and act on the implementation of the technology (hardware/infrastructure and software) to be used to complete the project, and then use it for these purposes.

Project Technology Management is the 12th performance domain and is made up of the following six processes:

- *Process 12.1*—**Plan Technology Management** (create a Technology Management Plan)
- *Process 12.2*—**Implement Technology** (execute the Technology Management Plan)
- *Process 12.3*—**Document Technology** (obtain or develop the project technology user documentation)
- *Process 12.4*—**Train on Technology** (deliver the project technology training)
- *Process 12.5*—**Report on Technology** (provide performance feedback on the use of the project technology)
- *Process 12.6*—**Integrate Technology** (integrate technology performance feedback) (as illustrated in Figure 4.2).

12. Project Technology Management
12.1 Plan Technology Management
12.2 Implement Technology
12.3 Document Technology
12.4 Train on Technology
12.5 Report on Technology
12.6 Integrate Technology

FIGURE 4.2
The six processes of Project Technology Management.

Stage Domain	Align	Plan	Execute	Check/Act	Confirm
Project Change Management		Plan Technology Management	Implement Technology Document Technology Train on Technology	Report on Tech Integrate Tech	

FIGURE 4.3

Table: the six PTechM processes across the five stages.

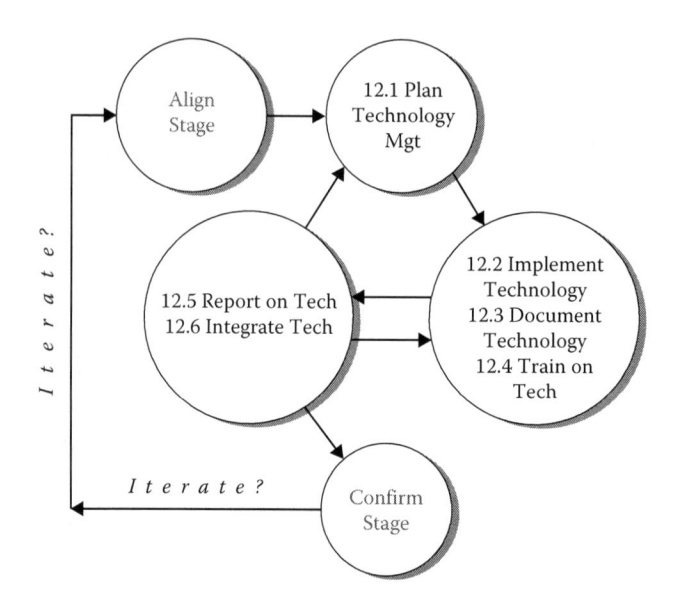

FIGURE 4.4

Graphic: the six PTechM processes across the five stages.

These six processes occur across three of the five Stages of the iterative Performance Improvement Life Cycle as depicted in Figures 4.3 and 4.4.

Now we present each of the six processes of Project Change Management.

PROCESS 12.1—PLAN TECHNOLOGY MANAGEMENT (create a Technology Management Plan)

As with each of the other Performance Domains, it's important to have a Management Plan as a component of the Project Management

Plan covering the vital aspects for how the technology should be used to plan, execute, and check/act on the project. Hence, for Plan Technology Management, this is the Technology Management Plan. Figure 4.5 illustrates the approximate chronological order of this process. (Note: See Chapters 6, 7, and 8 to see the order in which we believe each of these six Project Technology Management processes SHOULD be performed within each of the three Stages of our "Full" Approach.)

The Technology Management Plan is a subsidiary element or component of the Project Management Plan that describes which project technology resources (both hardware/infrastructure and software) will be used to complete the project and how it will be implemented, documented, trained on, reported upon, and integrated into the daily work routines of the performance improvement project team.

The sections of the Technology Management Plan should include, but not be limited to:

- The technology resources (both hardware/infrastructure and software) to be used to plan, execute, and check on the work of the project
- The process for implementing the Technology Management Plan
- The process for documenting the project technology resources
- The process for training on the project technology resources
- The process for reporting performance feedback on the project technology resources
- The process for integrating the performance feedback with future versions of the project technology documentation and training

Organizations that have a PMO with a portfolio of projects usually have an enterprise-wide hardware/infrastructure in place as well as a "standard" set of systems and software applications or a single system that is used for performing the four different types of work in organizations (see Figure I.1 in the Introduction of this book). In fact, our earlier

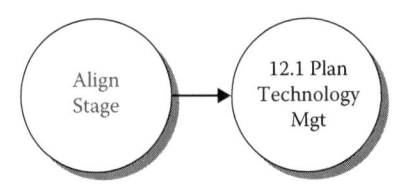

FIGURE 4.5
Plan Technology Management process.

book entitled *Effective Portfolio Management Systems* provides guidance in this regard.[3] Leveraging these enterprise-wide standards will reduce the amount of time the Project Manager of a single performance improvement project will need to devote to completing this process, as well as the others.

PROCESS 12.2—IMPLEMENT TECHNOLOGY (implement the provisions of the Technology Management Plan)

Once the Technology Management Plan has been created, it needs to guide the evaluation, selection, and implementation of the project technology and to check on their frequency of usage and level of adoption amongst the Project Team. The project technology selected to manage the project may include either a Commercial Off-The-Shelf (COTS) product or a customized application or system, or both, and they may be governed by the organization's PMO or IT organization (as illustrated in Figure 4.6).

PROCESS 12.3—DOCUMENT TECHNOLOGY (obtain or develop the project technology user documentation)

As mentioned in Process 12.2, the project technology selected to manage the project may be either a COTS product or customized application or system. If it is the former, then it should be configurable (but NOT customizable) and come with the vendor's pre-existing documentation and, therefore, should be provided along with the application or system when it is installed; if it's the latter case, then the documentation will need to be developed, often from scratch, to match the customized application or

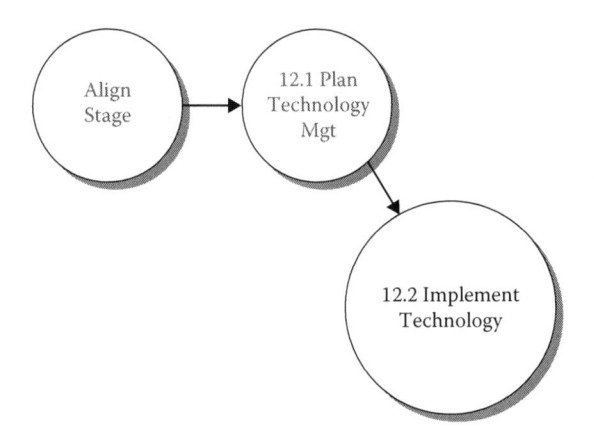

FIGURE 4.6
Implement Technology process.

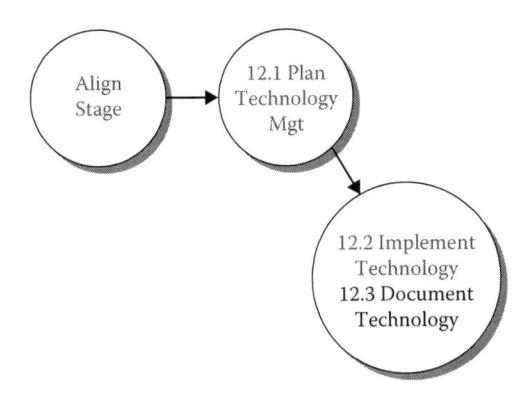

FIGURE 4.7
Document Technology process.

system. Here again, as before, the existence or creation of documentation may be governed by the organization's PMO (as illustrated in Figure 4.7).

PROCESS 12.4—TRAIN ON TECHNOLOGY (design, develop and deliver project technology training)

Once you have possession of the project technology documentation, it should be used to design and develop an appropriate training curriculum and, once approved, the delivery of training sessions to a cadre of "Power Users" or "Super Users" who should be carefully selected. (*Author's note*: These are stakeholders who have the reputation for being very savvy using technology and who will become mentors to those project technology users who need hands-on support.) This curriculum may be instructor-led, online, or a blend of the two, and the actual training can be done either synchronously, asynchronously, or a blend of the two. These training issues may be governed by the organization's PMO (Figure 4.8).

PROCESS 12.5—REPORT ON TECHNOLOGY (provide project technology performance feedback)

After the project technology has been implemented, documented, and had training delivered for it, it's important to report on how well the project technology is performing for the "Power Users" or "Super Users" and provide feedback to the key stakeholders. These stakeholders could include the Initiating Sponsor, the Sustaining Sponsor, the Customer, the Project Manager, the Performance Improvement Team, and the larger community of End Users. These usage feedback issues may be addressed and governed by the organization's PMO (Figure 4.9).

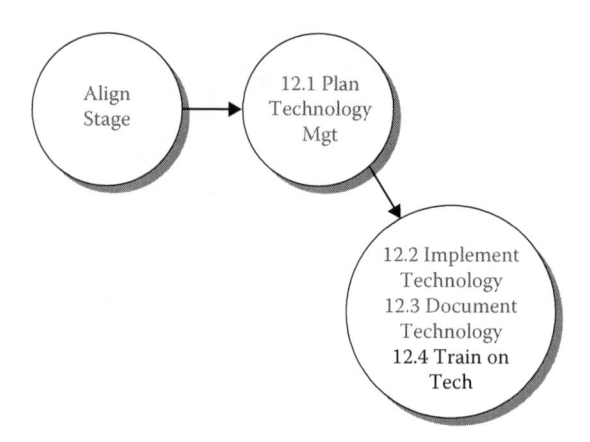

FIGURE 4.8
Train on Technology process.

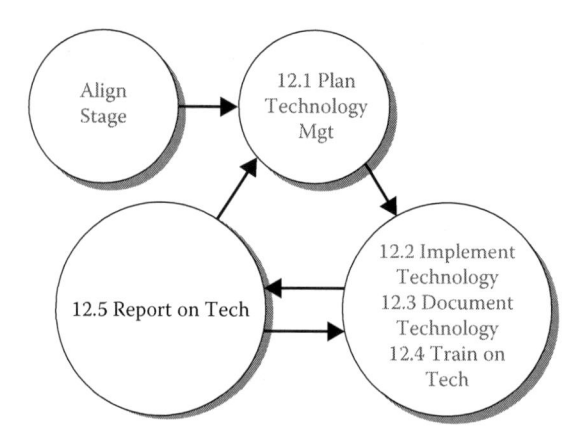

FIGURE 4.9
Report on Technology process.

PROCESS 12.6—INTEGRATE TECHNOLOGY (integrate project technology performance feedback)

Finally, you need to integrate the feedback that's been reported by the "Power Users" or "Super Users" on how well the project technology has been performing to date. Then, a decision needs to be made among three options:

1. Continue implementing and documenting the project technology, and designing, developing, and delivering training on the project technology as is.

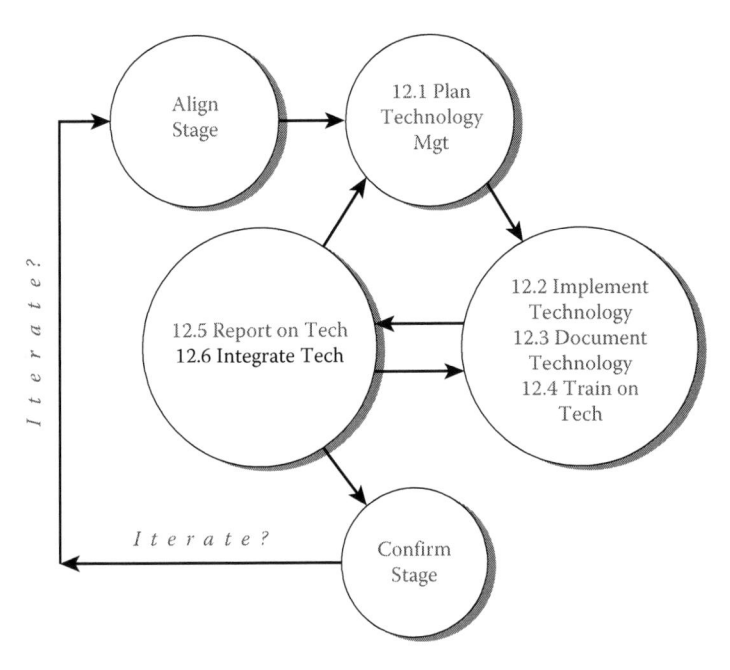

FIGURE 4.10
Integrate Technology process.

2. Revise the Technology Management Plan to change how you implement, document, and design/develop/deliver training for the project technology.
3. Pass on the results to the Confirm Stage to determine if the goals for using the project technology to complete the project have been met (as illustrated in Figure 4.10).

SUMMARY

By following the iterative PTechM model outlined in this chapter, organizations will be better positioned to maximize the return on investment of time and resources invested into the project technology being used to carry out their performance improvement projects. This results in a cadre of specially trained "Power Users" or "Super Users" who can lead the use of the technology among the larger community of users in their respective business units.

"To err is human; to really screw up requires project technology; and, to avoid as many screw-ups as possible, be sure to apply Project Technology Management!"

William S. Ruggles

REFERENCES

1. Project Management Institute, *The Project Management Body of Knowledge* (aka *PMBOK® Guide*), 6th Edition, p. 38, © 2017, Newtown Square, PA: Project Management Institute.
2. Christopher F. Voehl, Harrington H. James, and William S. Ruggles, *Effective Portfolio Management Systems*, p. xi, © 2016, Boca Raton, FL: CRC Press.
3. Ibid., Chapter 4, pp. 99–119.

PROJECT CASE STUDY EXAMPLE: ePROVIDE II

Here we resume describing how one of the co-authors addressed and applied Project Technology Management to "eProvide II", a User Provisioning and Identity Management Project in the IT Global Shared Services (GSS) Division of a Fortune 125 biopharmaceutical company that was first mentioned in this book's *Introduction* section (see the Summary at the end of the Introduction and Figures I.5 and I.6). At the time he was recruited to manage this project, user-provisioning implementations in small, medium, and large organizations were increasing due to regulatory compliance needs and enhancements in identity protection and security, role management, reporting, and industry support.

However, our focus in *this* particular part of the Project Case Study is *not* on the technical nature of the workflows and outcomes of the project deliverable, but on the "project technology resources" required to create the management plan and carry it out.

Here's how we applied the six processes of Project Technology Management on that project:

1. *Plan Technology Management* (create the Technology Management Plan)

    Since this was being performed within the Global Shared Services division of an international biopharmaceutical firm, there was already a strong commitment to IT-driven Standard Operating Procedures (SOPs) and various project-related Management Plans, including a template for the Technology Management Plan. While this rapidly maturing GSS organization did NOT have a Project Management Office (PMO) at that time, it DID have a Project Management Community of Practice (PMCoP) that was accumulating and sharing several reusable tools, techniques, and templates. All we had to do was complete the pre-existing template based on the unique requirements of the eProvide II Project. These technology resource categories included:

- **Computer Hardware/Network Environment:** Dell desktop and laptop PCs running Windows 10, Cisco LAN, Oracle, etc.
- **Planned Interfaces with Other Systems:** Active Directory, LDAP, HR, CFO, Documentum, etc.
- **Project Data Capture, Tracking, Repository, and Storage Policy:** MS Project, MS Project Server, MS Excel, MS Word, MS SharePoint client-server or cloud-based, etc.
- **Implementation and Support Methodology:** SDLC, Agile/Scrum, ITIL, ISO 21500, Stored Procedures for Oracle Account, etc.

The technology-driven processes we applied included those for implementing the Technology Management Plan; documenting, training on, and reporting performance feedback on the project technology resources; and integrating that performance feedback on a regular basis with future versions of the eProvide II Project Technology documentation and training.

2. *Implement Technology* (implement the provisions of the Technology Management Plan)

Once we created the initial version of the Technology Management Plan, we used it to guide the implementation of the project technology and to check on the frequency of usage and level of adoption amongst the Project Team. (Note: We did NOT have to evaluate and select the project technology since they had already been evaluated, selected, and standardized for us.) The project technology used to manage the project included mostly Commercial Off-The-Shelf (COTS) applications such as MS Project, MS Project Server, MS Excel, MS Word, MS SharePoint client-server (see Figure 4.11).

3. *Document Technology* (obtain or develop the project technology user documentation)

Since most of the project technologies used on the eProvide II Project were COTS applications, they came with the vendor's preexisting documentation. Therefore, each one was provided along with the application or system when it was installed. However, not all the members of the User Community were equally capable of mastering the project technology without outside support. That support was provided via instructor-led, small class training and private tutorials (see the next process: 4. Train on Technology).

4. *Train on Technology* (design, develop and deliver the project technology training)

As mentioned in the previous process, instructor-led, small class training and private tutorials were designed, developed, and delivered by the Learning Shared Services division to those members of the eProvide II User Community who requested it on a "just-in-time" basis.

FIGURE 4.11
Implement Technology on eProvide II.

Training on the eProvide II Project's technological environment was required, too. As a Contracted Subject Matter Expert, I had to successfully complete no fewer than 25 MANDATORY Information Security Policy (Technical) Training Modules within 60 days of my start date in order to continue serving as the Project Manager. These online modules (with Level of Mastery Tests at the end of each one) included the following: Wireless Communication Security; Computer System and Network Usage; Information Security Risk Assessment and Management; Third Party Network Connection Security; Compliance Assurance for Information Security; Authentication and Access Control; Remote Access Services Security; Corporate Privacy; GSS SDLC SOP; GSS Learning Life Cycle; Logging and Filing Procedures for Validation and SDLC Documents; IT Owner IMprove Training; Project Manager IMprove Training; and Sox 404 Awareness training, just to name several of them.

5. *Report on Technology* (provide technology performance feedback)

Periodic usage system audits of the project technology applications were performed. They confirmed our belief that both our eProvide II "Super Users" and the members of the larger eProvide II User Community were using the project technology appropriately and it was performing up to expectations.

6. *Integrate Technology* (integrate technology performance feedback)

At monthly intervals, we integrated the feedback that had been reported by the "Super Users" on how well the project technology had been performing to date. Then, decisions were made among the three typical options:

1. Continue implementing and documenting the project technology, and designing, developing, and delivering training on the project technology, as is.
2. Revise the Technology Management Plan to change how we implemented, documented, and designed/developed/delivered the project technology training.

3. Pass on the results to the Confirm Stage to determine if the goals for using the project technology to complete the project had been met.

In almost all the cases, it was Option 1 until the end of the project, when we applied Option 3. We had chosen wisely!

5

Stage #1: Align the Project

INTRODUCTION

Our contemporary framework for project management of performance improvement projects has five iterative Stages named **Align, Plan, Execute, Check/Act** and **Confirm** with the respective number of processes for each, in parentheses, inside each symbol as displayed in Figure 5.1.

Recall from Chapter 2 that our contemporary framework has done the following: (1) adds a new Stage to the traditional continual improvement cycle (called Align) *before* the Plan Stage; (2) combines two Stages (Check and Act) *after* the Plan Stage (called Check/Act); and (3) adds another new Stage at the end of the cycle (called Confirm).

This new framework of five Stages is also similar to the traditional project management framework, which has five Process Groups but with slightly different names for three of them (e.g., "Initiating" instead of Align, "Monitoring/Controlling" instead of Check/Act, and "Closing" instead of "Confirm"). However, the interactions of our five contemporary Stages differ significantly from the interactions of the five traditional Stages. (For illustrative purposes, compare Figure 2.6 for our contemporary framework to Figures 1.4 and 1.5 for their traditional framework.)

In this chapter and the next four chapters, we drill down and describe each of the five iterative Stages, one-by-one, in more detail. With them, you can apply project management to Performance Improvement Teams in an iterative and scalable way that matches the unique challenges and pragmatic struggles facing you in the 21st century using either a "Full," "Lean," or "Hybrid" Approach.

In this chapter, we cover the Align Stage, its three potential processes, its two Key (Pre-Project) Inputs, and its four Key Outputs. Then, as promised

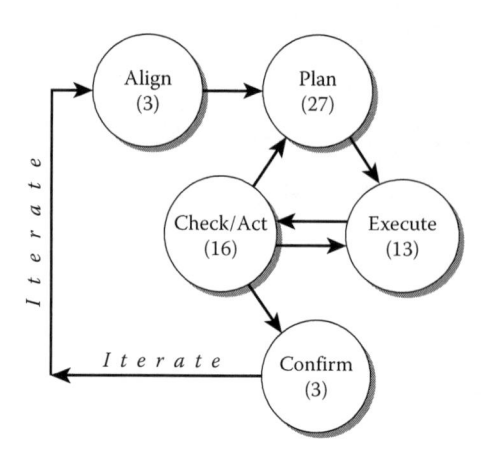

FIGURE 5.1
The five "APECC" Stages.

in the Introduction of this book, we close this chapter by relating how its contents apply to one of the two Case Studies: eProvide II.

ALIGN THE PROJECT

The primary perspective of this Stage is to ensure the strategic alignment of a new performance improvement project or a new Phase of an existing project by verifying its agreement with the organization's Mission Statement, and the terms and conditions of the project's Business Case, its Value Proposition, and its promised benefits. Secondarily, this Stage is focused on obtaining authorization to start the project or to iteratively continue on to perform the next Phase of the project. Next, it is in the Align Stage that a decision is made about which of the scalable project management Approaches—**Full**, **Lean**, or **Hybrid**—is recommended for the performance improvement project based on its relative Scope Size, and Priority within the organization. Finally, this Stage identifies the stakeholders for the performance improvement project. Since a Project Manager is needed to complete this Stage properly, it's assumed that one has been appointed *before* or will be appointed *concurrently with* the start of this Stage.

Projects and programs undertaken to exploit an "opportunity for improvement" or fulfill a business need that are *not* aligned with the organization's Mission Statement should *not* be considered for implementation unless there is a change to that Mission Statement that DOES incorporate them. Projects and programs that are not in line with the organization's culture should not be undertaken unless there is a project approved to bring the culture in line with the project or program. Projects and programs that are aligned with or support the organization's Mission Statement, but are *not* directly in line with the organization's strategic plan, can be considered candidates to become part of the organization's portfolio of active projects/programs. However, these latter projects and programs have a far lower chance of being funded and implemented than those that directly support the organization's strategic plan (Figure 5.2).

If your organization has a Project Portfolio System (PPS) or a Project Management Office (PMO), then it's likely that your performance improvement project would have already "run the gauntlet" of review and assessment by an objective third party (individual or group) as to its potential value. Then, too, it's likely that a variety of forms and templates will be available for you to use as "Key Inputs" in performing the "Key Steps" to produce the "Key Outputs" identified in the following paragraphs.

If such a PPS or PMO does NOT exist, then it will be important to be able to articulate where and how each project is aligned with the organization's Mission Statement. In addition, you'll either have to create the needed Align Stage forms and templates (e.g., Project Charter, Stakeholder Register, Change Agenda, etc.) from scratch or contact us to obtain them using the contact information provided at the end of the Preface of this book.

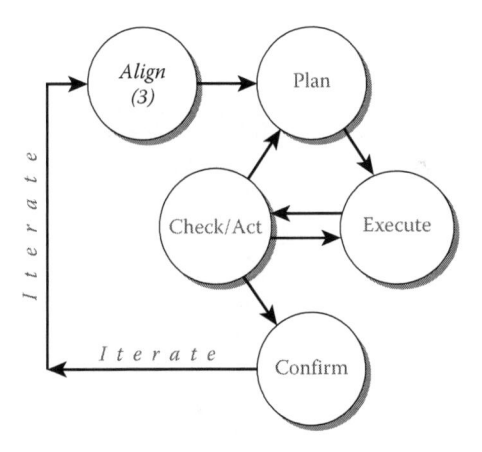

FIGURE 5.2
The Align Stage within APECC.

(*Authors' Note*: Since we embrace the 49 processes as they are described in Chapters 4 through 13 in the sixth edition of the *PMBOK® Guide*, we see NO need to repeat those descriptions in this book. Hence, we urge the reader to obtain a copy of this 2017 publication and refer to it if and when it is needed to better understand specific processes. The only exceptions are our two *new* Performance Domains—**Project Change Management** and **Project Technology Management**—and their 12 processes, which are documented in Chapters 3 and 4, respectively, in this book.)

PM4PITs Scalable Approach Options—Full, Lean, or Hybrid

The actual number of processes, Key Inputs, Key Steps, and Key Outputs applied can differ based on the scalable Approach to be used in managing it. This will usually depend upon the relative Scope Size and Priority Levels of each project: the "Full" versus the "Lean" versus a "Hybrid" Approach (a scaled-down "Full" Approach) as recommended in Figure 5.3:

Type #	Project scope	Project priority	Approach
1	Large	High	Full
2	Moderate	High	Full
3	Small	High	Hybrid
4	Large	Medium	Full
5	Moderate	Medium	Hybrid
6	Small	Medium	Lean
7	Large	Low	Hybrid
8	Moderate	Low	Lean
9	Small	Low	Lean

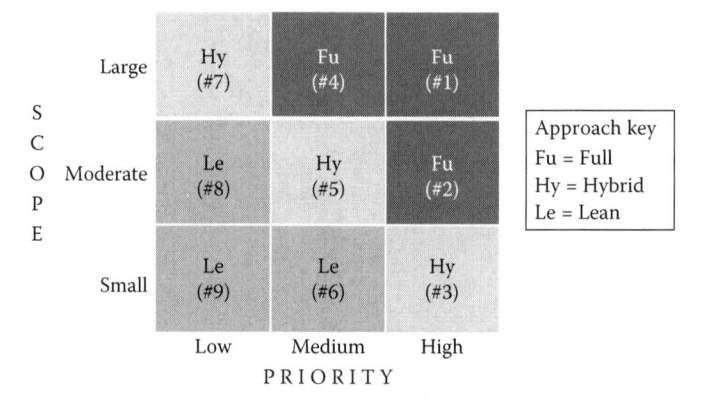

FIGURE 5.3
The Scalable Plan Stage Approach options.

Align Stage–Full Approach: Processes, Key Inputs, Key Steps, and Key Outputs

The Align Stage–Full has just three processes in three separate Performance Domains (Figure 5.4), but they are crucial processes indeed:

- **Develop Project Charter**[1]
- **Identify Stakeholders**[2]
- **Align for Change**

Key Inputs for the Align Stage–Full Approach

The two Key Inputs for the Align Stage–Full are:

- **Project Business Case:** A pre-project, documented evaluation of the potential impact an opportunity for improvement (OFI) has on the organization to determine if it is worthwhile investing the resources to take advantage of the OFI. It quantifies the Value Proposition or reason for justifying a potential project or program.
- **Project Benefits Management Plan:** Another pre-project, business document that describes the project's target benefits, how and when they will be delivered, and the mechanisms that should be in place to measure those benefits.

Key Iterative Steps for the Align Stage–Full Approach

Since it might not appear obvious to the inexperienced Project Manager, at first glance, where to start the Align Stage, permit us to suggest the following "typical" sequence of steps for completing it in an orderly yet iterative fashion as per Figure 5.5.

The three Key Steps for the Align Stage, as depicted in Figure 5.5 are:

- *Step #1* (Change Management): You need to start at the top and create a **Change Agenda** with a sense of urgency by performing the *Align for Change* process. Since the Project Manager may not be appointed yet, it's important that the Initiating Sponsor maintain ownership for completing this step until one is appointed or until a Sustaining Sponsor comes forward. It's important to use this

PM domains \ Stages	Align (3)	Plan (27)	Execute (13)	Check/Act (16)	Confirm (2)
1—Integration	*Develop Project Charter*	Develop Project Management Plan	Manage Project Work Manage Project Knowledge	Mon and Control Project Work Perform Integrated Change Control	Close Project/ Phase
2—Scope		Plan Scope Management Collect Requirements Define Scope Create Wbs		Validate Scope Control Scope	
3—Schedule		Plan Schedule Mgmt Define Activities Sequence Activities Estimate Activity Durations Develop Schedule		Control Schedule	
4—Cost		Plan Cost Management Estimate Costs Determine Budget		Control Costs	
5—Quality		Plan Quality Management	Manage Quality	Control Quality	
6—Resources		Plan Resource Mgmt Estimate Resources	Acq. Resources Develop Team Manage Team	Control Resources	
7— Communictions		Plan Communications Management	Manage Communicatns	Monitor Communications	
8—Risk		Plan Risk Management Identify Risks Perform Qualitative Risk Analysis Perform Quantitative Risk Analysis Plan Risk Responses	Implement Risk Responses	Monitor Risks	
9— Procurement		Plan Procurement Management	Conduct Procurements	Control Procurements	
10— Stakeholders	*Identify Stakeholders*	Plan Stakeholder Engagement	Manage Stakeholder Engagement	Monitor Stakeholder Engagement	
11—Change	*Align for Change*	Enroll for Change Plan Change Management	Manage Change	Check Change Management	Confirm Change
12— Technology		Plan Technology Management	Implement Technology Document Technology	Train Technology Report Tech Integrate Tech	

FIGURE 5.4
Align Stage–Full *and* its three processes.

top-down Approach to articulate a clear and compelling vision for where you're heading via the performance improvement project accompanied by a sense of urgency. The Project Manager and the entire Team need to be aligned with and committed to that vision, understand how disruptive the change created by the project is likely

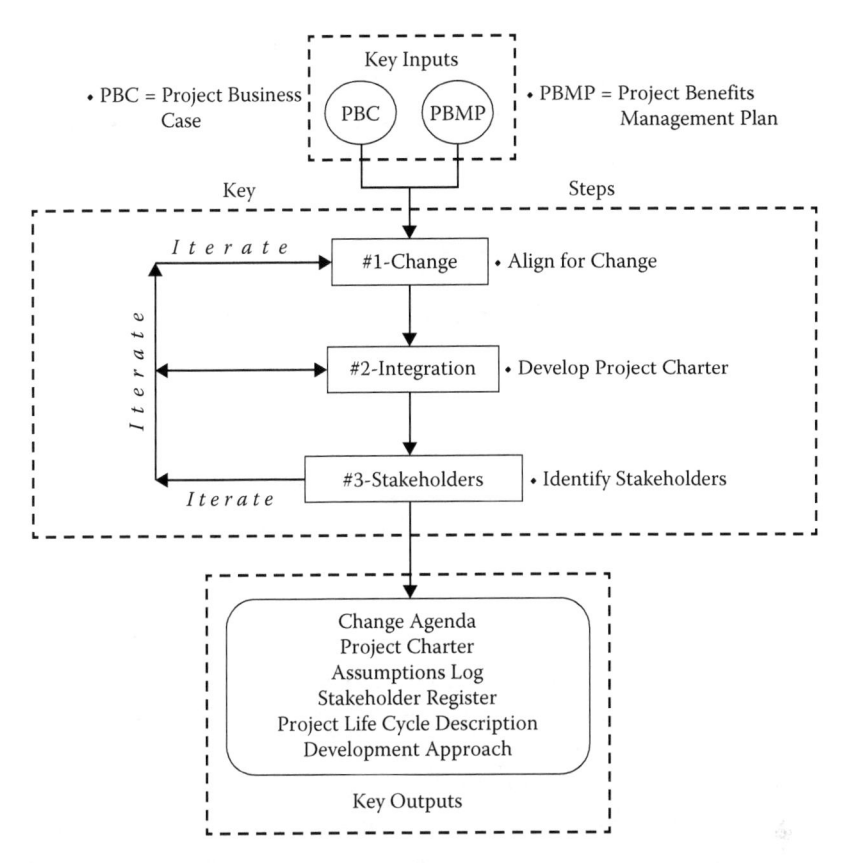

FIGURE 5.5
The Key Steps for the Align Stage–Full.

to be, and balance the ambition to change with the organization's capacity to absorb that change. See Chapter 3 in this book for more details on Project Change Management.

- **Step #2** (Integration Management): Now that you have created the Change Agenda to reflect the contents of the Project Business Case and the Project Benefits Management Plan, you should be ready to perform the **Develop Project Charter** process. Here again, if the Project Manager has yet to be appointed, it's up to the Initiating Sponsor to maintain ownership for completing this step until one is appointed or until a Sustaining Sponsor comes forward. The resulting **Project Charter** should be issued (signed off) by the Initiating Sponsor and, if one exists, the Sustaining Sponsor (Iterate, if needed.)[1]
- **Step #3** (Stakeholder Management): Now that you have created the Project Charter to go with the Change Agenda, you are ready

to create a **Stakeholder Register**, which is the primary output of the **Identify Stakeholders** process. This document contains information about the project's stakeholders in three categories: *Identification, Assessment,* and *Classification.* (Iterate, if needed.)[2]

Key Outputs for the Align Stage–Full Approach

The six Key Outputs for the Align Stage–Full Approach are:

- **Change Agenda:** A literal agenda of events that gets distributed as part of the Enrollment Plan in the next Stage to all prospective Change Agents (CAs). It sets the stage for meetings or conference calls held with the CAs to explain the purpose, scope, and milestones of the project; engage them in a dialog to enroll them in the effort; explain their roles; and show them how they shall become active participants in crafting the plan for change.
- **Project Charter:** The document, is issued by the Project Initiator or Sponsor that formally authorizes the existence of a project and provides the Project Manager in the Performing Organization with the authority to apply organizational resources on behalf of the Requesting Organization. It documents the high-level information of the project such as project purpose, measureable objectives and related success criteria, requirements, overall project risk, summary milestone schedule, pre-approved financial resources, key stakeholder list, and more. It establishes a partnership between the Performing Organization and the Requesting Organization but it is *not* a contract since there is no "consideration" paid.
- **Assumptions Log:** Since assumptions are the "seeds of risk" that are "sown" in the midst of uncertainty, this Project Document or Database records all those assumptions made throughout the iterative life cycle to be sure they are addressed during the team's rounds of project risk management, especially during the Plan Stage.
- **Stakeholder Register:** A Project Document that includes but is not limited to the identification and assessment information for, and the classification of, the project stakeholders.
- **Project Management Life Cycle Approach:** The scalable option selected for managing this project in an iterative fashion: **Full, Lean,** or **Hybrid.**

- **Development Approach:** The method used to create and elaborate the product, service, system, or result during the project management life cycle, such as a predictive (waterfall), iterative, adaptive, incremental, agile, or blended method.

Align Stage–Lean Approach: Processes, Key Inputs, Key Steps, and Key Outputs

The Align Stage-Lean Approach has even fewer processes—two—than the Full Approach, due to the relatively smaller Project Scope and Priority levels involved. Here are the two Key Inputs, two Key Steps, and five Key Outputs for the Align Stage–Lean Approach as per the Scalable Approach Options in Figure 5.6:

- Develop Project Charter[1]
- Identify Stakeholders[2]

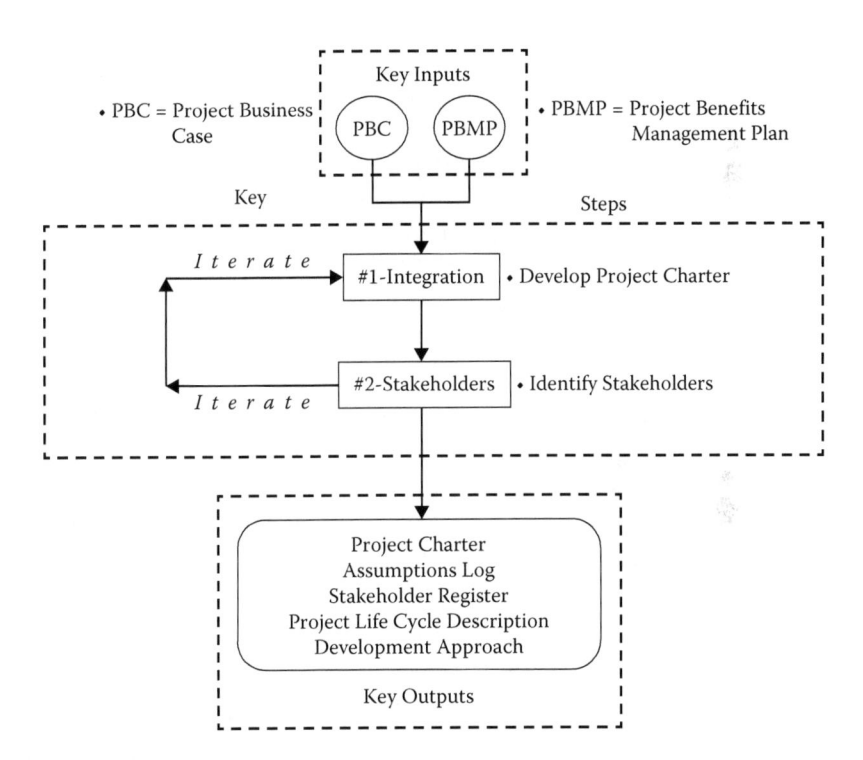

FIGURE 5.6
The Key Steps for the Align Stage–Lean.

Key Inputs for the Align Stage–Lean Approach

The two Key Inputs for the Align Stage–Lean are:

- **Project Business Case:** A pre-project, documented evaluation of the potential impact an opportunity for improvement (OFI) has on the organization to determine if it is worthwhile investing the resources to take advantage of the OFI. It quantifies the Value Proposition or reason for justifying a potential project or program.
- **Project Benefits Management Plan:** Another pre-project, business document that describes the project's target benefits, how and when they will be delivered, and the mechanisms that should be in place to measure those benefits.

Key Iterative Steps for the Align Stage–Lean Approach

Since it might not appear obvious to the inexperienced Project Manager, at first glance, where to start the Align Stage–Lean Approach for a smaller or lower-priority project, permit us to suggest the following "typical" sequence of steps for completing it in an orderly yet iterative fashion as per Figure 5.6.

When using a Lean Approach, there is a Key Assumption being made: because the relative Scope and Priority of the project will NOT have a significant impact on the Requesting or Receiving Organization, it will NOT require the use of Project Change Management. If this Key Assumption is determined to be FALSE, then the Lean Approach should NOT be used.

The two Key Steps for the Align Stage–Lean, as depicted in Figure 5.6, are:

- *Step #1* (Integration Management): Now, you should be ready to perform the **Develop Project Charter** process. Here again, if the Project Manager has yet to be appointed, it's up to the Initiating Sponsor to maintain ownership for completing this step until one is appointed or until a Sustaining Sponsor comes forward. The resulting **Project Charter** should be issued (signed off) by the Initiating Sponsor and, if one exists, the Sustaining Sponsor. (Iterate, if needed.)[1]
- *Step #2* (Stakeholder Management): Now that you have created the Project Charter, you are ready to create a **Stakeholder Register,** which is the primary output of the **Identify Stakeholders** process.

This document contains information about the project's stakeholders in three categories: *Identification*, *Assessment*, and *Classification*. (Iterate, if needed.)[2]

Key Outputs for the Align Stage–Lean Approach

The five Key Outputs for the Align Stage–Lean Approach are:

- **Project Charter:** The document, issued by the Project Sponsor, which formally authorizes the existence of a project and provides the Project Manager in the Performing Organization with the authority to apply organizational resources to project activities on behalf of the Requesting Organization. It establishes a partnership between the Performing Organization and the Requesting Organization but it is *not* a contract since there is no "consideration" paid.
- **Assumptions Log:** Since assumptions are the "seeds of risk" that are "sown" in the midst of uncertainty, this Project Document or Database records all those assumptions made throughout the iterative life cycle to be sure they are addressed during the team's rounds of project risk management, especially during the Plan Stage.
- **Stakeholder Register:** A Project Document that includes but is not limited to the identification and assessment information for, and the classification of, the project stakeholders.
- **Project Management Life Cycle Approach:** The scalable option selected for managing this project in an iterative fashion: **Full, Lean,** or **Hybrid**.
- **Development Approach:** The method used to create and elaborate the product, service, system, or result during the project management life cycle, such as a predictive (waterfall), iterative, adaptive, incremental, agile, or blended method.

Align Stage–Hybrid Approach: Processes, Key Inputs, Key Steps, and Key Outputs

The number of Processes, Key Inputs, Key Steps, and Key Outputs for the Align Stage–Hybrid Approach should be determined by the Project Manager and approved by the project's Initiating Sponsor based on the relative Scope and Priority levels of the project.

SUMMARY

In this chapter, we covered the Align Stage–Full Approach, its purpose, and its *three* potential processes: **Align for Change** (in Performance Domain #11—Project Change Management; **Develop Project Charter** (in Performance Domain #1—Project Integration Management); and **Identify Stakeholders** (in Performance Domain #10—Project Stakeholder Management), which should be performed sequentially yet iteratively. Then, we identified and described its two Key Inputs, three Key Steps, and six Key Outputs as illustrated in Figure 5.5.

We also did the same for the Align Stage–Lean Approach in which there are the same two Key Inputs, but only two Key Steps and five Key Outputs, as illustrated in Figure 5.6.

If a Hybrid Approach is determined to be appropriate for a given project, its Key Inputs, Key Steps, and Key Outputs should be identified and communicated to the Project Team BEFORE proceeding through each of the other four PM4PITs Stages which will be covered in Chapters 6–9.

Following the End Notes below, we close this chapter by relating how its contents apply to one of the two Case Studies: **eProvide II**.

REFERENCES

1. Project Management Institute, *The Project Management Body of Knowledge* (aka *PMBOK® Guide*), 6th Edition, pp. 75–83, © 2017, Newtown Square, PA: Project Management Institute.
2. Ibid., Chapter 13, pp. 507–515.

PROJECT CASE STUDY EXAMPLE: ePROVIDE II

Here we resume describing how one of the co-authors addressed and applied the principles, practices, and processes of the Align Stage to eProvide II—a User Provisioning and Identity Management Project in the IT Global Shared Services (GSS) Division of a Fortune 125 bio-pharmaceutical company that was first mentioned at the end of this book's Introduction (see the Summary at the end of the Introduction and Figures I.5. and I.6) and again at the end of Chapter 4, "Project Technology Management" (see Figure 4.11).

Here's how one of our co-authors led the Project Team in applying the three processes of the Align Stage–Full Approach to this project:

1. *Align for Change* (start from the top by communicating the benefits, challenges, risks and opportunities of the proposed change to gain the support of and buy-in from the CIO and her direct reports; create the eProvide II Project Change Agenda)

   Here is the short description of the project and the proposed change: "The eProvide II project consists of developing and implementing a new, enhanced, online provisioning solution that will replace eProvide and the R&D Account Request Form (ARF), providing the company with a single solution for submitting requests for hardware/networking, software, user IDs, and other services" (Figure 5.7).

   Since this was a "GSS Top Ten Project," most of the legwork for this process and the initial iteration of this Stage had already been done for me. All I had to do was acknowledge this "Top Ten List," continually reinforce it with the Team, recognize the resistance to adopting the new User Provisioning and Identity

---

Maximizing Business Value

The Top Ten focuses the efforts of Global Shared Services on key initiatives essential to our client's success. Selected against specific criteria, they provide diverse services for which our clients engage us while supporting and driving both the Company and Global Shared Services goals.

- Regional Order to Cash (ROtC)
- Product Lifecycle Management
- EMEA Business Intelligence
- TRECnet
- Global JET
- Caribbean SAP
- WorkSmart
- 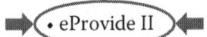 eProvide II
- Keystone (Call Center)
- GPACS

FIGURE 5.7
"GSS Top Ten" Projects poster.

Management system (especially among the soon-to-be-replaced Lotus Notes stakeholders), and prepare a Project Change Agenda that would address that resistance.

2. *Create the Project Charter* (Review the Business Case and the Benefits Management Plan; author a draft of the Project Charter and get it reviewed and signed off by the Project Sponsor)

Here again, since this was a "GSS Top Ten Project," I knew the Project Sponsor (the GSS Director of Provisioning and Identity Management) would be engaged with this project right from the start. I worked closely with his Associate Director (the GSS Technical Lead) who was a direct report to create the Project Charter, using a template that had been prepared and provided by the Project Management Community of Practice (PMCoP) Coordination Committee (Figure 5.8).

Here is the short description of the business value of the project that was embedded within that Project Charter we issued with the approval of the Project Sponsor and Technical Director:

> "Upon completion, eProvide II will replace the Lotus Notes version of eProvide and the R&D Account Request Form (ARF) with a customized, COTS application which will provide faster and easier processing of Service Requests via a simplified user

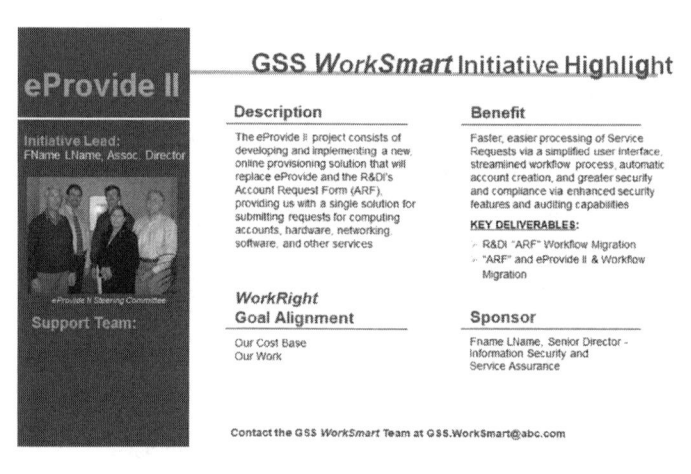

FIGURE 5.8
eProvide II WorkSmart poster.

> interface, streamlined workflow process, automatic account creation, and greater security and compliance via enhanced security features and auditing capabilities."

Since we had a section for "Assumptions and Constraints" embedded within the Project Charter, we did *not* create a separate "Assumptions Log." However, we DID create a "Risk Register" during the Plan Stage that was based on those assumptions and constraints and an "Open Issues Log" (we actually called it a "Parking Lot") during the Execute and Check/Act Stages for those risks and constraints that were realized throughout the life cycle.

3. *Identify the Project Stakeholders* (create the Stakeholder Register)

Our Stakeholder Register for this performance improvement project was referred to as the "eProvide II Project Team Rosters" with a total of some 257 stakeholders! Even if you were to eliminate duplicates, we still had more than 240 distinct stakeholders identified in that document! It was actually comprised of the following four Teams of Key Stakeholders (with their respective role and responsibilities within parentheses) totaling over 50 Key Stakeholders:

- **Executive Overview Team** (GSS CIO, VP, two Directors and two Associate Directors to attend a Monthly Status Meeting (except the CIO who would only attend quarterly) to receive a report on the overall status of the project, including budget and/or schedule variances; six members in all)
- **Steering Committee/Change Control Board** (GSS Director and four GSS Associate Directors to attend a meeting every three weeks to act as the final decision-makers pertaining to change requests and potential changes to the scope of the project; five members in all).
- **Core Team** (the Project Manager, Technical Lead, and all Project Leads representing the various Teams assigned to this project who meet every two weeks, as needed, to provide status and communicate project messages across the team; 19 members in all)

- **Extended Team** (all GSS Stakeholder Group representatives who would meet only occasionally with most status updates communicated via e-mail and other forms of electronic communications; 23 members in all)

In addition, there were two other Teams of "secondary" stakeholders who were identified up front but not utilized until toward the end of the project life cycle. These were the **User Acceptance Testing (UAT) Team** (comprised of 23 members) and the **Global Usability (GUI) Group** (comprised of 171 members) both of whom would become the "Key Clients" and "Super Users" of eProvide II. (See Figures I.5 and I.6 in the Introduction section of this book.)

We created and maintained a Stakeholder Register, which was a detailed spreadsheet comprised of each stakeholder and his/her contact information including name, company site, e-mail address, office phone #, and cell phone # and, after the Plan Stage–Full, we used it in conjunction with our Stakeholder Management Plan and Communications Management Plan.

6

Stage #2: Plan the Project

INTRODUCTION

In this chapter, we cover the **Plan** Stage in which, depending on the relative Scope and Priority of the project, the Project Manager may need to:

- Prepare a detailed "roadmap" for exploiting the *opportunity for improvement* (OFI) (e.g., goal, pain point, or problem, etc.) that was identified in the Align Stage and that is being addressed by this project.
- Refine its objectives.
- Determine the scope, quality, schedule, budget/resource, and risk constraints for the project.
- Create an Enrollment Plan to generate support for the Change Agenda.
- Create a Project Management Plan supported by assorted Project Documents with the current scope, schedule, and budget performance measurement baselines to be used to guide execution (see Figure 6.1).

It could involve as many as 27 processes in our contemporary framework for project management for performance improvement teams. As a result of the relatively high number of processes, as illustrated in Figure 6.2, the Plan Stage is potentially more complicated and time-consuming to complete than the other four Stages. For that reason, we encourage you to review our scaled-down Lean Approach described below.

The amount of time spent planning should be commensurate with the Scope and Priority of the project and the usefulness of the information developed by it. This means there are two extremes: on the one hand, "*Not*

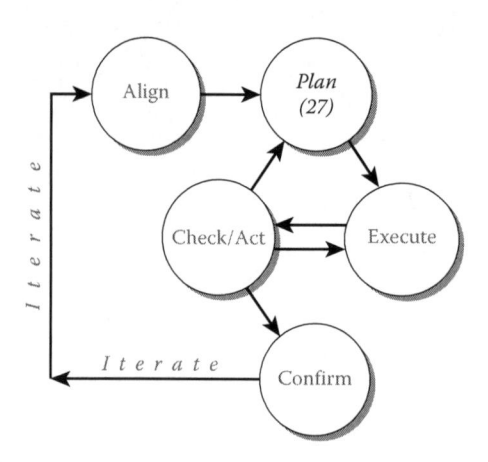

FIGURE 6.1
The Plan Stage within APECC.

Enough Planning" (better known as **the "Just Do It!" syndrome**) could miss identifying key constraints and assumptions that would increase the negative risk events (threats) associated with the project; on the other hand, *"Too Much Planning"* (better known as **the "Analysis Paralysis" syndrome**) could miss one or more "windows of opportunity" (positive risk events) associated with the project because it took too long.

The amount of time spent in the Plan Stage, like all the PM4PITs Stages, should be balanced. Planning is considered as much a "qualitative art" as it is an "exact science": two different teams could generate very different Project Management Plans for the same project, yet still be considered "compatible" with good project management practices.

Some inexperienced Project Managers complain that there's no need to waste so much time planning since all of the project variables are typically pre-negotiated or predetermined for them or imposed upon them in a top-down fashion before they even receive the assignment. This is particularly common in a project driven by a contract or purchase order that was nego-tiated by a "Sales Team" or "Account Team" of the "Seller" or "Provider" and passed on to its "Project Team" to develop.

The Project Team, in turn, is expected to do the best it can with what it's given in spite of the top-down constraints that may exist and the unidentified assumptions which may have been made, to produce results, not excuses. Even in-house projects that involve two different units play-ing the role of Buyer and Seller within the same organization have their fair share of these top-down, imposed project variables to deal with, too.

PM Domains \ Stages	Align (3)	*Plan* (27)	Execute (13)	Check/Act (16)	Confirm (2)
1—Integration	Develop Project Charter	*Develop Project Mgmt Plan*	Manage Project Work Manage Project Knowledge	Monitor and Control Project Work Perform Integrated Change Control	Close Project/ Phase
2—Scope		*Plan Scope Mgmt* *Collect Requirements* *Define Scope* *Create WBS*		Validate Scope Control Scope	
3—Schedule		*Plan Schedule Mgmt* *Define Activities* *Sequence Activities* *Estimate Activity Durtns* *Develop Schedule*		Control Schedule	
4—Cost		*Plan Cost Mgmt* *Estimate Costs* *Determine Budget*		Control Costs	
5—Quality		*Plan Quality Mgmt*	Manage Quality	Control Quality	
6—Resources		*Plan Resource Mgmt* *Estimate Resources*	Acq. Resources Develop Team Manage Team	Control Resources	
7—Communications		*Plan Communications Management*	Manage Communications	Monitor Communications	
8—Risk		*Plan Risk Management* *Identify Risks* *Perform Qualitative Risk Analysis* *Perform Quantitative Risk Analysis* *Plan Risk Responses*	Implement Risk Responses	Monitor Risks	
9—Procurement		*Plan Procurement Management*	Conduct Procurements	Control Procurements	
10—Stakeholders	Identify Stakeholders	*Plan Stakeholder Engagement*	Manage Stakeholder Engagement	Monitor Stakeholder Engagement	
11—Change	Align for Change	*Enroll for Change* *Plan Change Mgmt*	Manage Change	Check Change Management	Confirm Change
12—Technology		*Plan Technology Management*	Implement Technology Document Technology	Train Technology Report Tech Integrate Tech	

FIGURE 6.2
The Plan Stage and its 27 processes.

They're extremely common! So what should you do in such cases? Here's our advice.

In both cases, you've got to avoid the tendency to "blindly" accept these top-down, imposed deadlines, predetermined budgets, resource limitations, and open-ended lists of requirements, etc., and either validate or invalidate them with a solid "bottom-up" planning and estimating approach. This usually involves finding *at least* **one** of the constraints—Time, Cost, Scope, or Quality—that is "negotiable." Just remember: there is often great reluctance on the part of the "Buyer" or "Customer" to do so. Hence, you may have to accept more of the negative risk consequences than you'd prefer, unless you are prepared to walk away from the table (if you can).

If none of the above variables is negotiable, then you've got to identify, qualify, and quantify the resulting negative risk events (Threats) which exist, and prepare a Risk Register with a set of Risk Responses (also called a "Mitigation Strategy" or "Contingency Plan") for those events that exceed your acceptable threshold of risk, at least in terms of their likelihood of occurrence, impact on the project, and frequency of occurrence. To do otherwise would be foolish—the opposite of the due diligence planning concept we refer to in this chapter. Then, as promised in the Introduction of this book, we close this chapter by relating how its contents apply to one of the two Case Studies: Healthcare DCC.

PLAN THE PROJECT

The primary purpose of this Stage is to apply the project management life cycle and development approaches you identified as outputs from the Align Stage by using the level of due diligence that is appropriate for the Scope and Priority levels of the project. (*Authors' note*: Due diligence is defined as "the care that a reasonable person would exercise to avoid harm to other persons or their property.")

At this Stage, this due diligence should center upon: (1) preparing a detailed "roadmap" for exploiting the opportunity for improvement, fulfilling the goal, addressing the pain point, or solving the problem that was identified in the Align Stage and that is being addressed by this project; (2) determining the Scope, Quality, Schedule, Budget/Resource, and Risk constraints for the project and ensuring that they are defined together in a state of "balance" or "equilibrium" (don't forget the "Quadruple Constraint +1" introduced in Chapter 2!); (3) refining the project's objectives; and (4) creating a Project Management Plan with the current scope, schedule, and budget performance measurement baselines to be used to guide project execution and completion.

Other pertinent Performance Domains, depending on the relative scope and priority level of the project, could also include management plans covering Communications, Procurement, Stakeholders, Change, and Technology Management.

Plan Stage–Full: Processes, Key Inputs, Key Steps, and Key Outputs

The Plan Stage has 27 processes in all, as depicted in Figure 6.2, with *at least* **one** process in each of the 12 Performance Domains. If your organization has a Project Portfolio System (PPS) and/or a Project Management Office (PMO), then it's likely that a variety of forms and templates will be available for you to use as "Key Inputs" in performing the "Key Steps" to produce the "Key Outputs" identified in the following paragraphs. If you don't have a PPS or PMO, then you'll either have to create these forms and templates from scratch or contact us to obtain them using the information provided at the end of the Preface of this book.

(*Authors' note*: Since we embrace the 49 processes as they are described in Chapters 4 through 13 in the sixth edition of the *PMBOK® Guide*, we see NO need to repeat those descriptions in this book. Hence, we urge the reader to obtain a copy of this 2017 publication and refer to it if and when it is needed to better understand specific processes. The only exceptions are our two *new* Performance Domains—**Project Change Management** and **Project Technology Management**—and their 12 processes which are documented in Chapters 3 and 4, respectively, in this book.)

Here are the six Key Inputs, nine Key Steps, and three Key Outputs for the Plan Stage–"Full" Approach.

Key Inputs for the Plan Stage–Full Approach

Before you can expect to plan a performance improvement project both effectively and efficiently, you'll need to have *at least* the following six Key Inputs: **Change Agenda, Project Charter, Assumptions Log, Stakeholder Register, Project Life Cycle Description,** and **Development Approach,** which should have been created or identified during the Align Stage as "Key Outputs" in Chapter 5, as depicted in Figure 6.3.

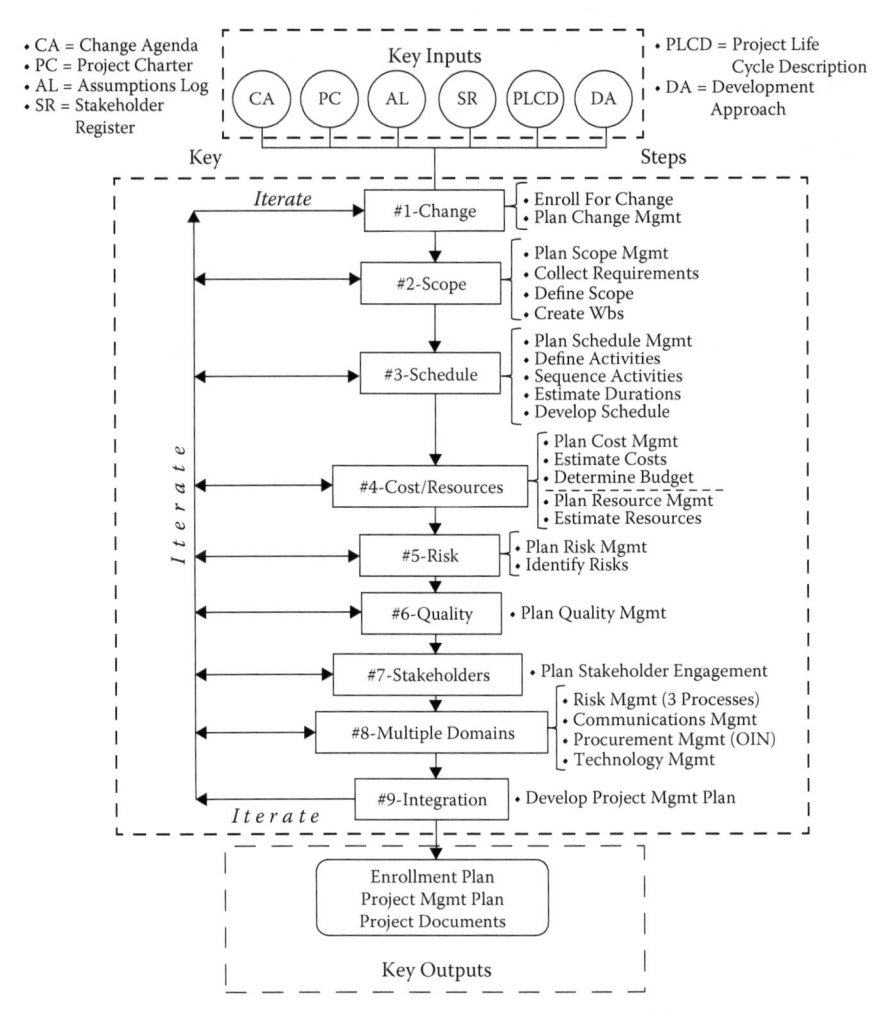

FIGURE 6.3
The nine Key Steps for the Plan Stage–Full.

Key Iterative Steps for the Plan Stage–Full Approach

Since it might appear intimidating to the inexperienced Project Manager where to start performing the Plan Stage at first glance, permit us to suggest the following "typical" sequence of steps for completing it in an orderly yet iterative fashion as per Figure 6.3. The nine Key Steps for the Plan Stage–Full Approach, as depicted in Figure 6.3, are:

> *Step #1* (Change Management): Use the Change Agenda you developed in the Align Stage to create the **Enrollment Plan** and the **Change**

Management Plan to provide further insights into the anticipated changes to the organization being caused by the performance improvement project. See Chapter 3 for more details. (Note: This strategic focus differs from the traditional project management framework, which is more concerned with changes to the performance improvement project caused by a realized opportunity or threat.)

Step #2 (Scope Management): Next, you need to plan the "WHAT" (Scope) before anything else. Using the "Key Outputs" from the Align Stage as "Key Inputs," perform the four Scope Management Planning processes (**Plan Scope Management**, **Collect Requirements**, **Define Scope**, and **Create WBS**), top down, in that order. (*Authors' note*: For those readers who aren't familiar with it, "WBS" stands "Work Breakdown Structure" and samples can be found in Figures 5.12, 5.13, and 5.14 on pages 158–160 of the *PMBOK® Guide*.)

Step #3 (Schedule–Management): Now you can address the following three related questions about your project: "WHEN"? "WHICH WORKFLOW"? "HOW LONG" (Schedule) should it take? As such, perform the five processes for Schedule Management (**Plan Schedule Management**, **Define Activities**, **Sequence Activities**, **Estimate Durations**, and **Develop Schedule**), top down, in that order. (Iterate, if needed.)

Step #4 (Cost/Resources Management): Next, you are ready to plan "HOW MUCH MONEY" (Budget) and "WHO AND HOW MANY RESOURCES" (Resources) the project will need. As such, perform the three Planning processes for Cost Management (**Plan Cost Management**, **Estimate Costs**, and **Determine Budget**) and the two Planning processes for Resource Management (**Plan Resource Management** and **Estimate Resources**) for your project. Perform each set of processes top down, in that order, to refine the budget. (Iterate, if needed.)

Step #5 (Risk Management): Now, you are ready to plan "HOW MUCH RISK" (both Opportunities and Threats) is involved in this undertaking. As such, perform the first two processes only (**Risk Management Planning** and **Identify Risks**) for Risk Management in a top-down sequence. (Iterate, if needed.)

Step #6 (Quality Management): Now, you can plan the "HOW" or "HOW WELL" (Quality) the work should be performed. As such, perform the single Quality Planning process (**Plan Quality Management**) at this time. (Iterate, if needed.)

Step #7: (Stakeholder Management): Now, you need to answer another key question for your performance improvement project: "WHO" needs to be engaged or committed to it to ensure its success? As such, perform the single Stakeholder Management Planning process (**Plan Stakeholder Engagement**) here. (Iterate, if needed.)

Step #8 (Multiple Domains): You should now be able to focus on the single planning process in each of these three performance domains: **Communications** (Plan Communications Management), **Procurement** (Plan Procurement Management, only if needed), and **Technology Management** (Plan Technology Management) in just about any order you prefer, as well as the three remaining **Risk Management Planning** processes (Qualitative Risk Analysis, Quantitative Risk Analysis, and Plan Risk Responses) in top-down order. (Iterate, if needed.)

Step #9 (Integration Management): Finally, you should leave the **Develop Project Management Plan** process in the Integration Management Performance Domain until the LAST iterative step of the Plan Stage. Why? Because the Project Management Plan will contain MOST of the other "Key Outputs" identified below as its subsidiary components, accompanied by several Project Documents, and the Enrollment Plan. (Iterate, if needed.)

Key Outputs for the Plan Stage–Full Approach

The three Key Outputs for the Plan Stage–Full Approach—the Project Management Plan, the Project Documents, and the Enrollment Plan—are comprised of the following components:

1. **Enrollment Plan**
2. **Project Management Plan**
   - Management Plans for: Scope, Requirements, Schedule, Cost, Quality, Resources, Communications, Risk, Procurement, Stakeholder (Engagement Plan), Change, Technology, and Configuration
   - Performance Measurement Baselines
     - Scope Baseline (includes the Project Scope Statement, Work Breakdown Structure, WBS Dictionary)
     - Schedule Baseline
     - Cost Baseline (Budget)

3. Project Documents

- Requirements Documentation, Requirements Traceability Matrix, Project Schedule (both Time-Driven and Resource-Driven), Change Request Form, Cost Estimates, Quality Metrics, Responsibility Assignment Matrix (RAM), Change Log, Issue Log, Assumptions Log, Resource Breakdown Structure, and Risk Register

Plan Stage–Lean: Processes, Key Inputs, Key Steps, and Key Outputs

Here are the five Key Inputs, one Key Step, and two Key Outputs for the "Plan Stage– Lean" Approach as per the Scalable Approach Options in Figure 5.6:

Key Inputs for the Plan Stage–Lean Approach

Before you can expect to plan a "Lean" performance improvement project properly, you'll need to have *at least* the following five Key Inputs: **Project Charter, Assumptions Log, Stakeholder Register, Project Life Cycle Description**, and **Development Approach,** which should have been created or identified during the Align Stage as Key Outputs in Chapter 5, as depicted in Figure 6.4.

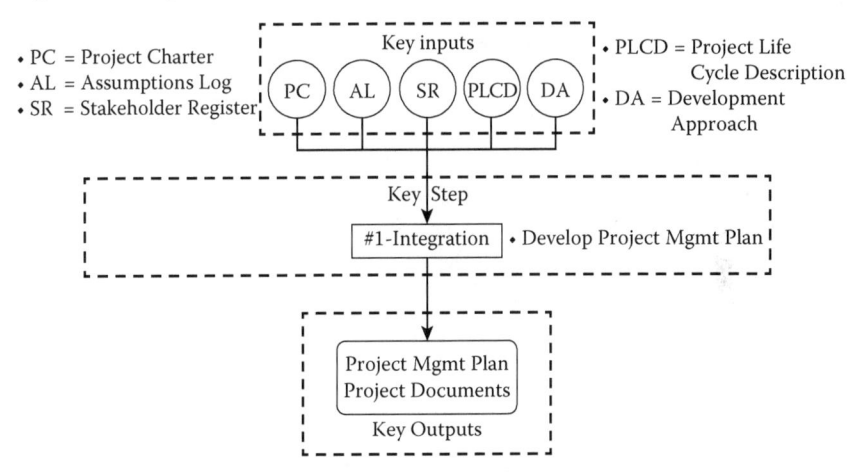

FIGURE 6.4
The Key Step for the Plan Stage–Lean.

Key Iterative Steps for the Plan Stage–Lean Approach

Using the "Lean" version of this contemporary framework for planning is much less intimidating to the inexperienced Project Manager since there is only ONE step as per Figure 6.4.

The one Key Step for the Plan Stage–Lean, as depicted in Figure 6.4, is:

Step #1 (Integration Management): You should focus solely on the **Develop Project Management Plan** process in the Integration Management Performance Domain as the only step of the Plan Stage–Lean Approach. Why? Because the Project Management Plan will contain MOST of the other Key Outputs identified below as its subsidiary components as well as several Project Documents as Key Outputs.

Key Outputs for the Plan Stage–Lean Approach

The two Key Outputs for the Plan Stage–Lean Approach—the Project Management Plan and the Project Documents—are comprised of the following components:

1. **Project Management Plan:** Could include the Management Plans for: Scope, Requirements, Schedule, Cost, Quality, Resources, Communications, Risk, Procurement, Stakeholder (Engagement Plan), Change, Technology, and Configuration; the Scope Baseline (includes the Project Scope Statement, Work Breakdown Structure, WBS Dictionary), the Schedule Baseline, and Cost Baseline (Budget), which serve as the Performance Measurement Baselines.
2. **Project Documents:** Could include the Requirements Documentation, the Requirements Traceability Matrix, the Project Schedule (both Time-Driven and Resource-Driven), the Change Request Form, the Cost Estimates, the Quality Metrics to be used, the Responsibility Assignment Matrix (RAM), the Change Log, the Issue Log, the Assumptions Log, the Resource Breakdown Structure, and the Risk Register.

Plan Stage–Hybrid Approach: Processes, Key Inputs, Key Steps, and Key Outputs

The number of Processes, Key Inputs, Key Steps, and Key Outputs for the Plan Stage–Hybrid Approach should be determined by the Project

Manager and approved by the Requesting Organization and the Project Sponsor, based on the relative Scope and Priority levels of the project.

SUMMARY

In this chapter, we covered the Plan Stage, its purpose and its scalable framework, its 27 Processes, and its Key Inputs, Key Steps, and Key Outputs for both the "Full" and "Lean" Approaches.

We close this chapter by relating how its contents apply to one of the two Case Studies: **Healthcare DCC**.

PROJECT CASE STUDY EXAMPLE: HEALTHCARE DCC

Here we resume describing how one of the co-authors addressed the challenges and applied the principles, practices, and processes for the Plan Stage to the Healthcare DCC program (see the Summary and Figures 1.3 and 1.4) and a portfolio of 59 projects contained in the Center's Master Going-Forward Plan (see Figure 3.13) and Work Plans/Timelines (see Figures 3.13 through 3.17).

As mentioned at the end of Chapter 3, we finished up the initial calendar year on a very positive note:

- Twelve of the 17 projects were completed before the two-week holiday break and the other five were scheduled to be finished before the end of the following month (Figure 6.5).

However, shortly after returning from the holiday break, the Acting P.I. held a special meeting of the entire DCC staff. He announced that he had been notified by our funding source that its Senior Leadership Team had decided that our end-of-calendar-year performance was sufficient to merit being awarded another renewable grant the following year, commencing in July of that year. There was a collective sigh of relief coming from the assembly of DCC staff in that large room and you could almost feel the stress level and, along with it, the sense of urgency drop through the floor. NO MORE PRESSURE TO PERFORM!

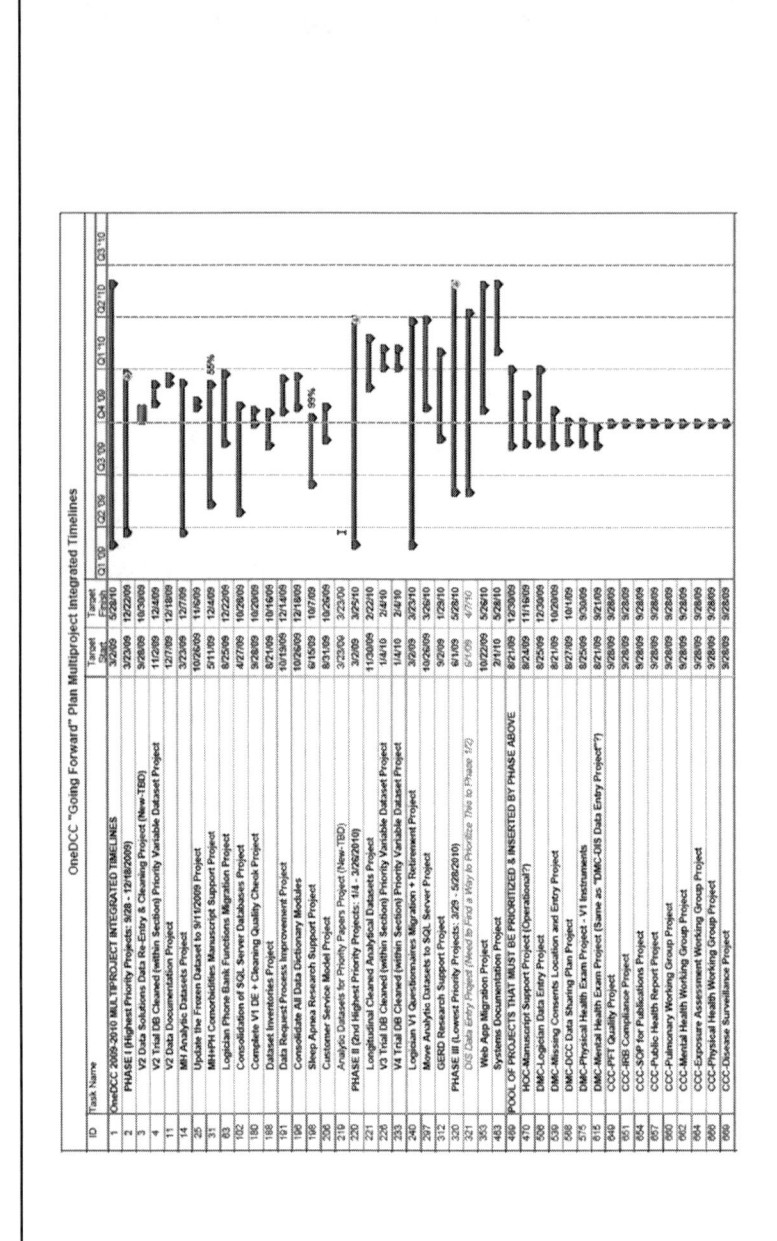

FIGURE 6.5
Healthcare DCC Plan Stage: multiproject work plans and timelines.

While we still had our OneDCC Going Forward Plan and the Multiproject Integrated Timelines for Phases II and III to go with it, they weren't worth the screens on which they were displayed or the paper on which they were printed out. The focus had almost instantaneously shifted from RESULTS and DELIVERABLES to JOB SECURITY and RELATIONSHIPS once again. It was time to iterate and "REPLAN"!

7

Stage #3: Execute the Project Work

INTRODUCTION

In this chapter, we focus on the processes needed to carry out the Enrollment Plan, the Project Management Plan, and the Project Documents for each agreed-upon periodic interval (weekly, bi-weekly, monthly, etc.) to fully meet the project requirements in a timely and cost-effective manner. We believe the "Full" Approach to this Stage is "anchored" by four processes found in our contemporary framework, including the Project Change Management ("**Manage Change**"), Project Resource Management ("**Acquire Resources**"), and Project Integration Management ("**Direct and Manage Project Work**" and "**Manage Project Knowledge**") Performance Domains, with support coming from nine processes in seven other Performance Domains as illustrated in Figure 7.1.

As with the Align and Plan Stages already covered, we have identified three scalable Approaches—*Full*, *Lean* or, *Hybrid*—which can be optionally applied to this Stage depending on the project's relative Scope and Priority levels. These are documented below.

We close this chapter, as with the previous chapters, by relating how its contents apply to one of the two Case Studies introduced previously: eProvide II.

EXECUTE THE PROJECT WORK

The primary purpose of this Stage is to carry out the detailed "roadmap" including the Project Management Plan and the Project Documents that were created in the Plan Stage to perform the project work; to maintain

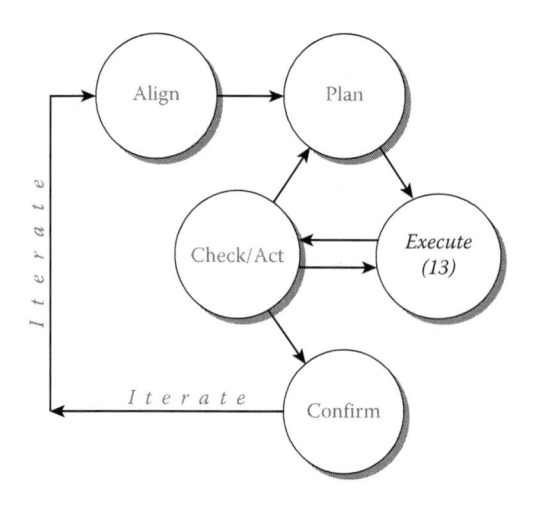

FIGURE 7.1
The Execute Stage within APECC.

the state of "balance" or "equilibrium" among the project's constraints; to implement the Enrollment Plan to ensure that the change(s) being *caused by* the project are being embraced by the stakeholders; and to ensure timely and cost-effective project completion.

This Stage has 13 processes as illustrated in Figure 7.2.

Execute Stage–Full: Processes, Key Inputs, Key Steps, and Key Outputs

The Execute Stage has 13 processes in all, as depicted in Figure 7.2. If your organization has a Project Portfolio System (PPS) and/or a Project Management Office (PMO), then it's likely that a variety of forms and templates will be available for you to use as "Key Inputs" to go through the "Key Steps" to produce the "Key Outputs" identified below. If you don't have one, then you'll either have to create these forms and templates from scratch or contact us to obtain them using the contact information provided at the end of the Preface of this book.

(*Authors' note*: Since we embrace the 49 processes as they are described in Chapters 4 through 13 in the sixth edition of the *PMBOK® Guide*, we see NO need to repeat those descriptions in this book. Hence, we urge the reader to obtain a copy of this 2017 publication and refer to it if and when it is needed to better understand specific processes. The only exceptions are our two *new* Performance Domains—**Project Change**

Stages / PM Domains	Align (3)	Plan (27)	*Execute (13)*	Check (16)	Confirm (2)
1—Integration	Develop Project Charter	Develop Project Mgmt Plan	*Manage Project Work* *Manage Project Knowledge*	Monitor and Control Project Work Perform Integrated Change Control	Close Project/ Phase
2—Scope		Plan Scope Management Collect Requirements Define Scope Create WBS		Validate Scope Control Scope	
3—Schedule		Plan Schedule Mgmt Define Activities Sequence Activities Estimate Activity Durations Develop Schedule		Control Schedule	
4—Cost		Plan Cost Management Estimate Costs Determine Budget		Control Costs	
5—Quality		Plan Quality Management	*Manage Quality*	Control Quality	
6—Resources		Plan Resource Mgmt Estimate Resources	*Acquire Resources* *Develop Team* *Manage Team*	Control Resources	
7—Communications		Plan Communications Management	*Manage Communications*	Monitor Communications	
8—Risk		Plan Risk Management Identify Risks Perform Qualitative Risk Analysis Perform Quantitative Risk Analysis Plan Risk Responses	*Implement Risk Responses*	Monitor Risks	
9—Procurement		Plan Procurement Management	*Conduct Procurements*	Control Procurements	
10—Stakeholders	Identify Stakeholders	Plan Stakeholder Engagement	*Manage Stakeholder Engagement*	Monitor Stakeholder Engagement	
11—Change	Align for Change	Enroll For Change Plan Change Mgmt	*Manage Change*	Check Change Management	Confirm Change
12—Technology		Plan Technology Management	*Implement Technology* *Document Technology*	Train Technology Report Tech Integrate Tech	

FIGURE 7.2
The Execute Stage and its 13 processes.

Management and **Project Technology Management**—and their 12 processes, which are documented in Chapters 3 and 4, respectively, in this book.)

Here are the three Key Inputs, six Key Steps, and 16 Key Outputs for the Execute Stage–Full Approach.

Key Inputs for the Execute Stage–Full Approach

Before you can expect to execute a performance improvement project that meets the "Full" Approach criteria both effectively and efficiently, you'll need to have *at least* the following three Key Inputs, which should have been created or identified during the "Plan" Stage as per Figure 7.3:

- **Enrollment Plan** (see also the Chapter 6 "Key Outputs").
- **Project Management Plan** (see also the Chapter 6 "Key Outputs").
- **Project Documents** (see also the Chapter 6 "Key Outputs").

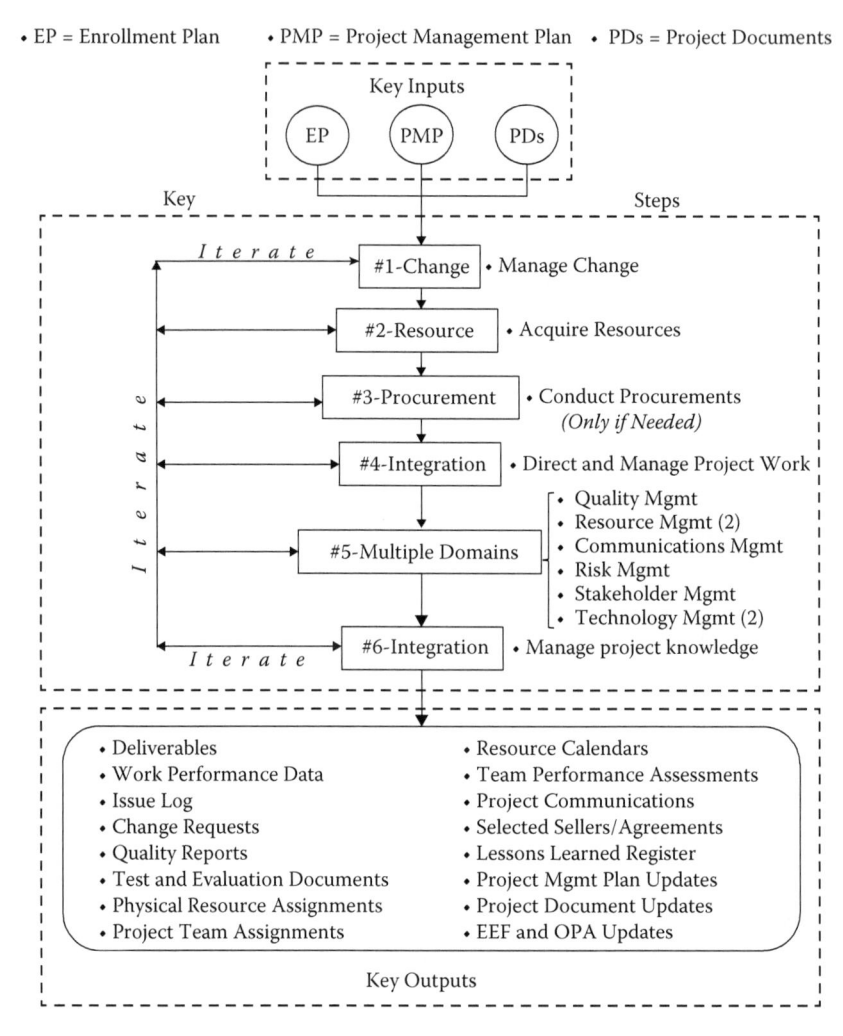

FIGURE 7.3

The Key Steps for the Execute Stage.

Key Iterative Steps for the Execute Stage–Full Approach

Since it might appear intimidating to the inexperienced Project Manager where to begin performing the Execute Stage, at first glance, permit us to suggest the following "typical" sequence of steps for completing it in an orderly yet iterative fashion as depicted in Figure 7.3.

The six Key Steps for the Execute Stage–Full Approach, as depicted in Figure 7.3 are:

Step #1 (Change Management): Use the **Enrollment Plan** Key Output from the Plan Stage as the Key Input to perform the **Manage Change** process. See Chapter 3 for more details.

Step #2 (Resource Management): Next, since you can't perform work without the needed resources who have the appropriate skill sets, you need to perform the **Acquire Resources** process in the Resource Management Performance Domain. (Iterate, if needed.)

Step #3 (Procurement Management): If you find that there aren't enough internal staff (quantitatively) who have the matching skill sets (qualitatively), you will need to look outside the Performing Organization for them. This would require performing the **Conduct Procurements** process in the Procurement Management Performance Domain. (Iterate, if needed.)

Step #4 (Integration Management): Now, using the items included in the **Project Management Plan** and **Project Documents** (two of the Key Outputs from the Plan Stage) as Key Inputs, perform the **Direct and Manage Project Work** process.

Step #5 (Multiple Domains): You should now be able to focus on the eight planning processes in each of these six performance domains: **Quality** (Plan Quality Management), **Resources** (Develop Team and Manage Team), **Communications** (Manage Communications), **Risk** (Implement Risk Responses), **Stakeholders** (Manage Stakeholder Engagement), and **Technology Management** (Implement Technology and Document Technology) in just about any order you prefer in top-down fashion. (Iterate, if needed.)

Step #6 (Integration Management): Finally, you should now be able to accumulate all of the Outputs and Lessons Learned accrued during the present iteration of this Stage by performing the second Execute process in the Integration Management Performance Domain: **Manage Project Knowledge**. (Iterate, if needed.)

Key Outputs for the Execute Stage–Full Approach

The 17 Key Outputs for the Execute Stage–Full Approach as depicted in Figure 7.3 are the following:

- Deliverables, Work Performance Data, Issue Log, Change Requests, Quality Reports, Test and Evaluation Documents, Physical Resource Assignments, Project Team Assignments, Resource Calendars, Team Performance Assessments, Project Communications, Selected Sellers/ Agreements, Lessons Learned Register, Project Management Plan Updates, Project Documents Updates, Enterprise Environmental Factors Updates, and Organizational Process Assets Updates.

Execute Stage–Lean: Processes, Key Inputs, Key Steps, and Key Outputs

Here are the two Key Inputs, three Key Steps, and eight Key Outputs for the Execute Stage–Lean Approach as per the Scalable Approach Options in Figure 5.3.

Key Inputs for the Execute Stage–Lean Approach

Before you can expect to execute a performance improvement project that meets the "Lean" Approach criteria both effectively and efficiently, you'll need to have *at least* the following two Key Inputs, which should have been created or identified during the Plan Stage as per Figure 7.4.

- **Project Management Plan** (see also the Chapter 6 "Key Outputs").
- **Project Documents** (see also the Chapter 6 "Key Outputs").

Key Iterative Steps for the Execute Stage–Lean Approach

Using the "Lean" version of this contemporary framework for executing the work is much less intimidating to the inexperienced Project Manager since there are only THREE steps. Since it might appear intimidating to the inexperienced Project Manager where to begin performing the Execute Stage, at first glance, permit us to suggest the following "typical" sequence of steps for completing it in an orderly yet iterative fashion as depicted in Figure 7.4.

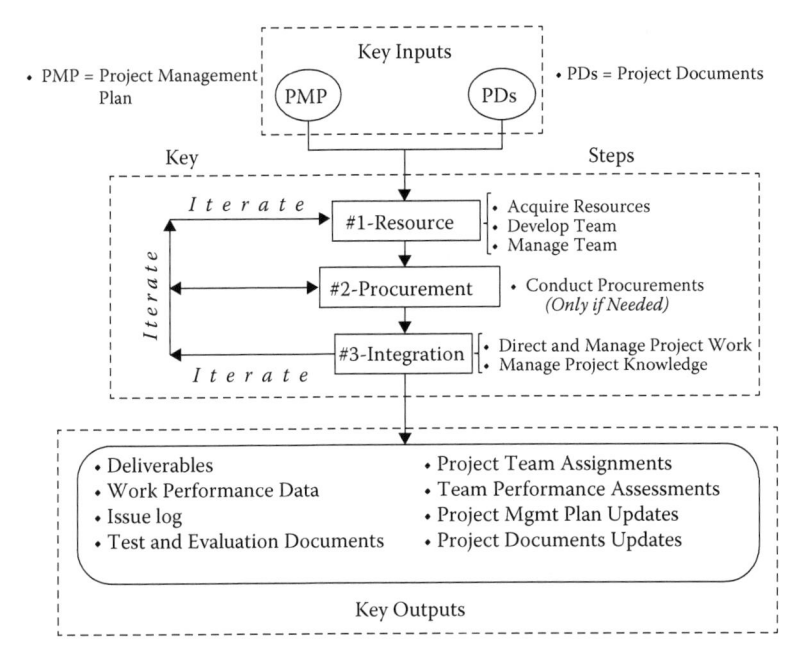

FIGURE 7.4
The Key Steps for the Execute Stage–Lean Approach.

The three Key Steps for the Execute Stage–Lean Approach, as depicted in Figure 7.4 are:

Step #1 (Resources Management): Since you can't perform work without the needed resources who have the appropriate skill sets, you need to perform the **Acquire Resources** process in the Resource Management Performance Domain. (Iterate, if needed.)

Step #2 (Procurement Management): If you find that there aren't enough internal staff (quantitatively) who have the matching skill sets (qualitatively), you will need to look outside the Performing Organization for them. This would require performing the **Conduct Procurements** process in the Procurement Management performance domain. (Iterate, if needed.)

Step #3 (Integration Management): Now, using the items included in the **Project Management Plan** and **Project Documents** (two of the Key Outputs from the Plan Stage) as Key Inputs, perform the **Direct and Manage Project Work** and **Manage Project Knowledge** processes. (Iterate, if needed.)

Key Outputs for the Execute Stage–Lean Approach

The eight potential Key Outputs for the Execute Stage–Lean Approach, as depicted in Figure 7.4, are the following:

- Deliverables, Work Performance Data, Issue Log, Test and Evaluation Documents, Project Team Assignments, Team Performance Assessments, Project Management Plan Updates, and Project Document Updates.

Execute Stage–Hybrid Approach: Processes, Key Inputs, Key Steps, and Key Outputs

The number of Processes, Key Inputs, Key Steps, and Key Outputs for the Execute Stage–Hybrid Approach should be determined by the Project Manager and approved by the Project Stakeholders (especially the Requesting Organization and the Project Sponsor) based on the relative Scope and Priority levels of the Project.

SUMMARY

In this chapter, we covered the "Execute" Stage, its purpose and scalable framework, its 13 Processes, its Key Inputs, its Key Steps, and its Key Outputs for both the "Full" and "Lean" Approaches. We focused on the "Execute" processes needed to carry out the Enrollment Plan ("Full" Approach only) and the Project Management Plan using various Project Documents for each agreed-upon periodic interval (weekly, bi-weekly, monthly, etc.) to fully meet the project requirements in a timely and cost-effective manner.

The Execute Stage is the most underappreciated Stage or "unsung hero" in our contemporary PM4PITs because most people think of the "Product Development" or "System Development" activities when they hear "Execute" or "Execution." (*Note*: Other more experienced people who've been made "Project Scapegoats" in the past think of facing a "firing squad" or "kangaroo court" when you mention the word "Execute" or "Execution." You can't blame them for that, now, can you?!)

This Stage is the one most directly affected by the project application area or industry in that the product or result of the performance improvement project is actually modified or enhanced here. As such, the vast majority of the project budget will be expended in performing and completing this Stage.

We close this chapter by relating how its contents apply to one of the two Case Studies: **eProvide II**.

PROJECT CASE STUDY EXAMPLE: ePROVIDE II

Here we resume describing how one of the co-authors addressed the challenges and applied the principles, practices, and processes for the Execute Stage of the pharmaceutical eProvide II project. This private-sector Case Study was originally presented at the end of the Introduction of this book (see the Summary and Figures I.5 and I.6 at the end of the Introduction), at the end of Chapter 4 (see Figure 4.11), and at the end of Chapter 5 (see Figures 5.7 and 5.8).

Execute the Work

Believe it or not, this Stage is where the co-author believes he demonstrated his unique value to the large pharmaceutical company that brought him in as a "hired gun" or "mercenary" Project Manager. While there are 13 processes in this Stage, he performed NINE of them on a daily or weekly basis. Here's where he focused his attention.

Develop Team, Manage Team, and Manage Communications Processes

We not only had a large team of employees—more than 240 in all—involved at one time or another, we also had a small team of outside consultants working with us. As such, the Project Manager had to look for ways to facilitate our passing through the first three Stages of the "Tuckman Model of Team Development"[1]: Forming, Storming, and Norming" as quickly as possible so that we could become a "High Performing" team.

All meetings:

SHOULD:
- Start/end on-time
- Focus on data and process
- Close with action plans or assigned follow-up items
- Include all appropriate people

SHOULD HAVE:
- A known agenda (which is then followed)
- Clearly stated goals and objectives

SHALL HONOR:
- One speaker at a time
- No interruptions
- The opportunity for all to participate
- All inputs and observations
- Consensus decisions

SHALL BE DEVOID OF:
- Personal attacks
- Side-bar conversations
- Hidden agendas

INDIVIDUALS SHALL:
- BE PUNCTUAL
- Listen carefully
- Contribute and participate
- Support consensus decisions
- Have fun
- Display the "7 Core Behaviors"
- Reinforce the "Company Pledge"

FIGURE 7.5
eProvide II Meeting Code of Conduct.

So, the Project Manager made sure he made time to "manage-by-wandering-around" (MBWA) on the main campus where most of the employees and contractors involved on his project had their offices. He also made sure that the company's "7 Core Behaviors" signs were posted in each conference room where our periodic Project Status Meetings were held. These 7 Core Behaviors were: *Leads Strategically, Builds Alignment, Communicates Directly, Drives Performance, Collaborates, Energizes Others,* and *Develops Other People.*

Finally, he also made sure that everyone received a copy of the "Meeting Code of Conduct," which he enforced relentlessly but discretely (see Figure 7.5).

Conduct Procurements

The Vendor team was a small group of employees of a local Consulting Partner for the Vendor of the COTS system for which the pharmaceutical

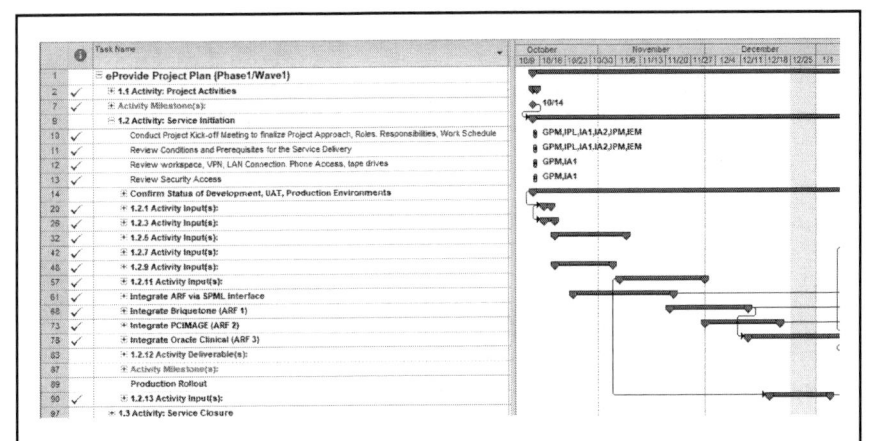

FIGURE 7.6
eProvide II project timeline rolled up.

company had purchased a global license. In addition to assisting in the deployment of the User Provisioning and Identity Management COTS system, the Vendor had also been subcontracted for their applications development and programing expertise to provide a carefully specified set of customized changes to the system before the contract was completed. The Associate Director for Provisioning Services made it clear to the Project Manager that he expected HIM to manage the Vendor and its extended team. This extended team included a "virtual" Project Manager who would show up on-site one day per week; an Architect, who would show up one week per month or when special meetings were held; and three developers who were stationed on-site every day.

Manage Project Work, Manage Project Knowledge, and Manage Change

Fortunately, since we had taken the necessary time to perform all of the Plan Stage processes, we had several documents—in both digital and hard-copy formats—available to help us guide the Execute Stage. One of these documents was a timeline created using MS Project that we updated weekly (see Figure 7.6). These allowed us to manage the work activities, the knowledge acquisition, and any changes that could impact the larger organization.

Another output that was helpful in managing these three areas was the "Key Interim Deliverables Status Chart" (see Figure 7.7).

Deliverable	Architect Says	A.D.-P.S. Says
A system that is ready for ARF application migration.	DONE (except for the production environment which is still OPEN).	We still need to close the production environment.
Connectors for LDAP and Oracle are configured using SSL.	DONE	DONE
The Provisioning application repository and a new Administrative tool.	The Repository is DONE, but the new Administrative Tool is still OPEN.	We still need to close the Administrative Tool.
The IA&M User Repository populated w/ Directory entries.	DONE (except for execution of the E.D. recon which is still OPEN).	We still need to close the E.D. recon.
Three company default workflows.	DONE	DONE
An approval and Administrative Console.	DONE	DONE
Two Oracle apps w/ automated account creation via IA&M Tool in UAT and DEV.	PC Image is DONE, BroqTine is still OPEN.	We still need to close BroqTine.

FIGURE 7.7
eProvide II Project Key Interim Deliverable Status Chart.

Yet another Plan Stage output that was helpful in managing these three areas was the color-coded "eProvide II/ARF Action Item Spreadsheet," which became our "Project Punch List" as we got closer and closer to our interim deliverable deadlines (see Figure 7.8).

Manage Quality

Another set of Plan Stage outputs we had created were consistent with the company's Quality Policy (which we posted several times): a Requirements Management Plan and a Requirements Document (that we used for the user interface, too), a Qualification Plan, and GSS System Development Life Cycle and Best Practices documents.

Another "tool" we built on our own well into the project's life cycle was an "Open Issues Parking Lot" spreadsheet. We started using it on

A	B	C	D	E	F	G	H
ID#	Action Item	Who	Category	Due Date	Comments	Notes	Status
001							
002							
003							
004							
005							
006							
007							
008							
009							
010							

FIGURE 7.8
eProvide II Project Action Items/Punch List.

A	B	C	D	E	F	G	H
ID#	Open Issue	Status	Responsible	Open Date	Update Date	Comments/Alternatives	Priority
OI-1							
OI-2							
OI-3							
OI-4							
OI-5							
OI-6							
OI-7							
OI-8							
OI-9							
OI-10							

FIGURE 7.9
eProvide II Project Open Issues Parking Lot.

a weekly basis but ended up using it on a daily basis as we got closer and closer to interim go-live dates, as displayed in Figure 7.9.

Manage Stakeholder Engagement

Since this "Top Ten Project" was going to impact, either directly or indirectly, everyone in this global company, we knew we had to "sweat the small stuff." So we contracted with a well-known GUI architect to be sure the new eProvide II system would have a user-friendly interface. We held focus group meetings with our Global Usability Group comprised of over 170 representatives from different stakeholder groups throughout the company and, then, we did something rare: WE LISTENED TO THEM. In fact, our "eProvide II Battle Cry" was "*3 Clicks to Anywhere!*" and it ended up being VERY popular with our stakeholders.

We also held periodic Town Halls at various company sites where we performed "Live Pilot Demonstrations" of its functionality to in-person groups of other stakeholders to check on our progress and obtain their feedback, too.

CASE STUDY END NOTE

1. B.W. Tuckman, 'Developmental sequence in small groups', © 1965, *Psychological Bulletin,* Vol. 63, pp. 384–399.

8

Stage #4: Check/Act on the Latest Performance Data

INTRODUCTION

In this chapter, we focus on the processes needed to check on the latest performance data for each agreed-upon periodic interval (weekly, bi-weekly, monthly, etc.) to determine the performance improvement project's progress to date and forecast the expected values for being able to successfully meet the project objectives in a timely and cost-effective manner, or not.

This Stage is actually a combination of BOTH "Check" AND "Act" because of the iterative nature of the conditional interactions between three different Stages: (1) between "Check/Act" and "Execute"; (2) between "Check/Act" and "Plan"; and (3) between "Check/Act" and "Confirm." Therefore, each one of those conditional interactions would be "acting on" the latest performance data. Here's how it should work:

You should review and analyze the integrated performance data from the Execution Stage—especially for the Scope, Schedule, and Budget Performance Baselines—for the specified periodic interval and identify what the data indicates. Then, depending on the comparative results, take one of the following three actions ("Act On"), as illustrated in Figure 8.1:

1. *If* the actual performance to date, including the value of the work performed, was consistent with the planned performance ("NO variance") and no changes are needed, *then* iterate back to **Execute** to perform the work scheduled for the next periodic interval (see Chapter 7); *else*

2. *If* the actual performance-to-date, including the value of the work performed, was *not* consistent with the planned performance

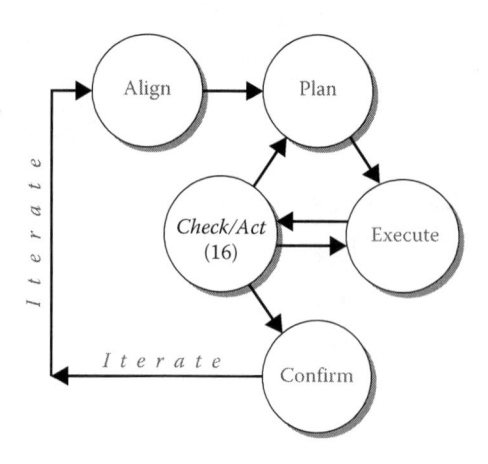

FIGURE 8.1
The Check/Act Stage within APECC.

("variance") and one or more changes are needed, **then** a Change Request needs to be generated and submitted to the appropriate entity (i.e., Change Control Board). If the Change Request is approved, proceed to (Re-)**Plan** to make the approved change in the work scheduled for the next periodic interval (and other future periodic intervals, if needed). **Then**, after performing this replanning, proceed to the **Execute Stage** to perform the revised work schedule for the next periodic interval (see Chapter 6); **else**

3. **If** the actual performance to date, including the value of the work performed, was consistent with the planned performance ("NO variance"), **no** changes are needed, and there is **no** more work scheduled for the current phase of the project life cycle, OR **if** the project needs to be terminated prematurely before all of the work has been executed ("NO more work to be executed"), **then** proceed to the **Confirm Stage** to perform its Key Steps (see Chapter 9).

So, now you should understand the reason we use the word "iterative" to describe our contemporary framework for PM4PITs and how we have embedded "Act" or "Act On" within the "Check" Stage as described in Paragraphs 1, 2, and 3 above.

As with the Align, Plan, and Execute Stages already covered, we have identified three scalable Approaches—*Full, Lean* or *Hybrid*—which can be optionally applied to this Stage depending on the project's relative Scope and Priority levels as per Figure 5.3. These options are documented below.

We close this chapter, as with the previous chapters, by relating how its contents apply to one of the two Case Studies introduced previously: Healthcare DCC.

CHECK THE LATEST PERFORMANCE TO DATE

The primary purpose of this Stage is to review and analyze project performance and the results of execution for that particular periodic interval and, based on what has occurred, take the necessary action steps. This Stage has up to 16 processes as illustrated in Figure 8.2.

Check/Act Stage–Full: Processes, Key Inputs, Key Steps, and Key Outputs

The Check/Act Stage–Full Approach has 16 processes in all, as depicted in Figure 8.2. If your organization has a Project Portfolio System (PPS) and/or a Project Management Office (PMO), then it's likely that a variety of forms and templates will be available for you to use as "Key Inputs" in performing the "Key Steps" to produce the "Key Outputs" identified in the following paragraphs. If you don't have a PPS or PMO, then you'll either have to create these forms and templates from scratch or contact us to obtain them using the contact information provided at the end of the Preface of this book.

(*Authors' note*: Since we embrace the 49 processes as they are described in Chapters 4 through 13 in the sixth edition of the *PMBOK® Guide*, we see NO need to repeat those descriptions in this book. Hence, we urge the reader to obtain a copy of this 2017 publication and refer to it if and when it is needed to better understand specific processes. The only exceptions are our two *new* Performance Domains—**Project Change Management** and **Project Technology Management**—and their 12 processes, which are documented in Chapters 3 and 4, respectively, in this book.)

Here are the nine Key Inputs, seven Key Steps, and five Key Outputs for the Check/Act Stage–Full" Approach, as illustrated in Figure 8.3.

Key Inputs for the Check/Act Stage–Full Approach

Before you can expect to effectively and efficiently check the periodic performance of an improvement project that meets the "Full" Approach criteria, you'll need to have the following nine Key Inputs, which should have

Stages PM Domains	Align (3)	Plan (27)	Execute (13)	Check/Act (16)	Confirm (2)
1—Integration	Develop Project Charter	Develop Project Mgmt Plan	Manage Project Work Manage Project Knowledge	*Monitor and Control Project Work* *Perform Integrated Change Control*	Close Project/ Phase
2—Scope		Plan Scope Management Collect Requirements Define Scope Create WBS		*Validate Scope* *Control Scope*	
3—Schedule		Plan Schedule Mgmt Define Activities Sequence Activities Estimate Activity Durations Develop Schedule		*Control Schedule*	
4—Cost		Plan Cost Management Estimate Costs Determine Budget		*Control Costs*	
5—Quality		Plan Quality Management	Manage Quality	*Control Quality*	
6—Resources		Plan Resource Mgmt Estimate Resources	Acq. Resources Develop Team Manage Team	*Control Resources*	
7—Communications		Plan Communications Management	Manage Communications	*Monitor Communications*	
8—Risk		Plan Risk Management Identify Risks Perform Qualitative Risk Analysis Perform Quantitative Risk Analysis Plan Risk Responses	Implement Risk Responses	*Monitor Risks*	
9—Procurement		Plan Procurement Management	Conduct Procurements	*Control Procurements*	
10—Stakeholders	Identify Stakeholders	Plan Stakeholder Engagement	Manage Stakeholder Engagement	*Monitor Stakeholder Engagement*	
11—Change	Align For Change	Enroll For Change Plan Change Mgmt	Manage Change	*Check Change Management*	Confirm Change
12—Technology		Plan Technology Management	Implement Technology Document Technology	*Train Tech* *Report Tech* *Integrate Tech*	

FIGURE 8.2
The Check/Act Stage and its 16 processes.

been created or identified during the Execute Stage as per Figure 8.3 (see also Chapter 7's "Key Outputs"):

- Project Management Plan, Project Documents, Agreements, Work Performance Reports, Work Performance Data, Change Requests, Work Performance Info, Verified Deliverables, and Approved Change Requests.

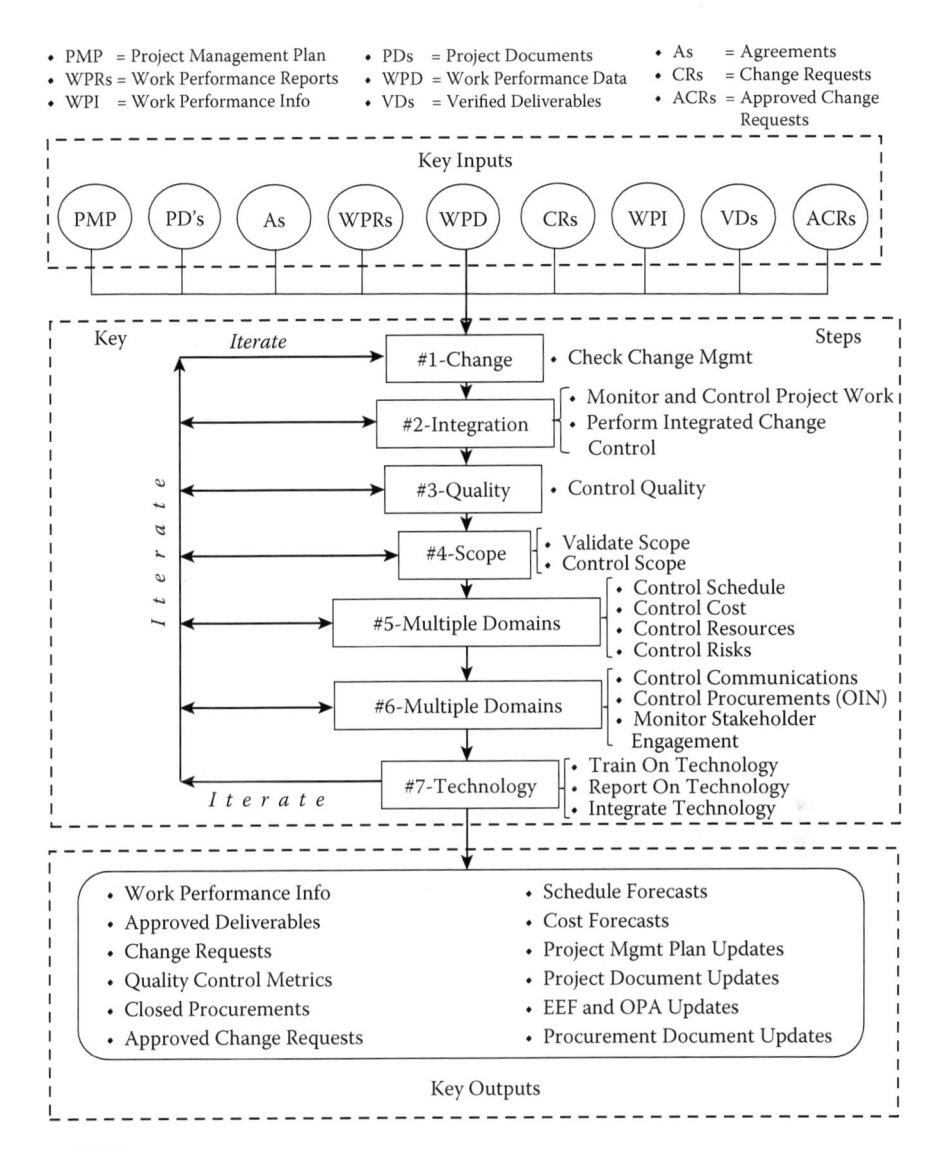

- PMP = Project Management Plan
- WPRs = Work Performance Reports
- WPI = Work Performance Info
- PDs = Project Documents
- WPD = Work Performance Data
- VDs = Verified Deliverables
- As = Agreements
- CRs = Change Requests
- ACRs = Approved Change Requests

FIGURE 8.3

The Key Steps for the Check/Act Stage–Full.

Key Iterative Steps for the Check/Act Stage–Full Approach

Since it might appear intimidating, at first glance, to the inexperienced Project Manager how and when to perform the Check/Act Stage properly, permit us to suggest the following "typical" sequence of steps for completing it in an orderly yet iterative fashion as per Figure 8.3.

The seven Key Steps for the Check/Act Stage, as depicted in Figure 8.3, are:

Step #1 (Change Management): Use the **Change Management Plan** and **Change Agenda** as the Key Inputs to perform the **Check Change Management** process. The focus here, in contrast to Step #2 below (**change(s) on or to the project**), should be on any **change(s) being caused by** the project. See Chapter 3 for more details. (Iterate, if needed.)

Step #2 (Integration Management): Next, you can use the **Configuration Management Plan** (and part of the **Change Management Plan**) to perform the **Monitor and Control Project Work** process followed by the **Perform Integrated Change Control** process. The focus here, in contrast to Step #1 above, should be on:

1. Tracking, reviewing, and reporting the overall progress of the project to meet the performance objectives defined in the Project Management Plan
2. Reviewing all change requests, approving (or denying) them, managing changes to the project's deliverables, documents, and Management Plans, and communicating all change-related decisions to the appropriate stakeholders. (Iterate, if needed.)

Step #3 (Quality Management): Next, you can perform the **Control Quality** process to inspect the deliverables produced for *correctness* to declare them "*Verified Deliverables*" or not. These, in turn, will be reviewed for *acceptance* in the **Validate Scope** (see Step #4) process to declare them "*Approved Deliverables*" or not. (Iterate, if needed.)

Step #4 (Scope Management): Now, use the "*Verified Deliverables*" output from the **Control Quality** process to perform the two Scope Management processes: **Validate Scope** and **Control Scope**. (Iterate, if needed.)

Step #5 (Multiple Domains): You should now be able to focus on the four Check/Act Stage processes in each of these four Performance Domains: **Schedule** (Control Schedule), **Cost** (Control Cost), **Resources** (Control Resources), and **Risk Management** (Control Risks) in just about any order you prefer. (Iterate, if needed.)

Step #6 (Multiple Domains): next, you should be able to focus on the three Check/Act Stage processes in each of these three Performance Domains: **Communications** (Control Communications),

Procurement (Control Procurements, only if needed), and **Stakeholder Management** (Monitor Stakeholder Engagement) in just about any order you prefer. (Iterate, if needed.)

Step #7 (Technology Management): Finally, you should now be able to perform the three **Technology Management** processes (Train on Technology, Report on Technology, and Integrate Technology). (Iterate, if needed.)

Key Outputs for Check/Act Stage–Full Approach

The Key Outputs for the Check/Act Stage Full Approach, as depicted in Figure 8.3, are the following:

- Work Performance Info, Approved Deliverables, Change Requests, Quality Control Metrics, Closed Procurements, Approved Change Requests, Project Team Assignments, Schedule Forecasts, Cost Forecasts, Project Management Plan Updates, Project Document Updates, Enterprise Environmental Factors Updates, Organizational Process Assets Updates, and Procurement Document Updates

Check/Act Stage-Lean: Processes, Key Inputs, Key Steps, and Key Outputs

Here are the nine Key Inputs, one Key Step, and five Key Outputs for the Check/Act Stage–Lean Approach:

Key Inputs for the Check/Act Stage–Lean Approach

Before you can expect to effectively and efficiently check and act on the periodic performance of an improvement project that meets the "Lean" Approach criteria both effectively and efficiently, you'll need to have the following nine Key Inputs, which should have been created or identified during the Execute Stage as per Figure 8.4 (see also Chapter 7's "Key Outputs"):

- Project Management Plan, Project Documents, Agreements, Work Performance Reports, Work Performance Data, Change Requests, Work Performance Info, Verified Deliverables, and Approved Change Requests.

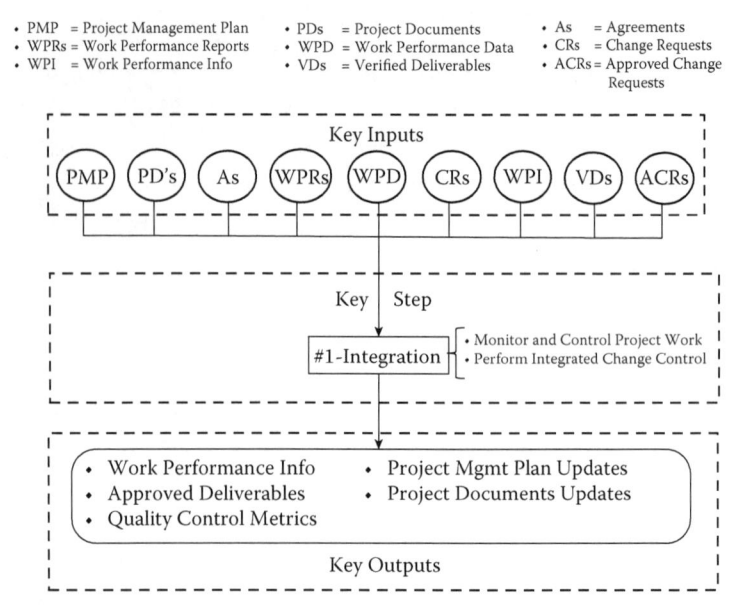

FIGURE 8.4
The Key Steps for the Check/Act Stage–Lean Approach.

Key Iterative Steps for the Check/Act Stage–Lean Approach

Since it might appear intimidating, at first glance, to the inexperienced Project Manager how and when to perform the Check/Act Stage properly, permit us to suggest the following "typical" sequence of steps for completing it in an orderly yet iterative fashion as per Figure 8.3.

The one Key Step for the Check/Act Stage, as depicted in Figure 8.4, is:

> *Step #1* (Integration Management): Use the **Configuration Management Plan** (and part of the **Change Management Plan**) to perform the **Monitor and Control Project Work** process followed by the **Perform Integrated Change Control** process. The focus here should be on:
>
> 1. Tracking, reviewing, and reporting the overall progress of the project to meet the performance objectives defined in the Project Management Plan
> 2. Reviewing all change requests, approving (or denying) them, managing changes to the project's deliverables, documents, and Management Plans, and communicating all change-related decisions to the appropriate stakeholders. (Iterate, if needed.)

Key Outputs for Check/Act Stage–Lean Approach

The Key Outputs for the Check/Act Stage–Lean Approach, as depicted in Figure 8.4, are the following:

- Work Performance Info, Approved Deliverables, Quality Control Metrics, Project Management Updates, and Project Documents Updates

Check/Act Stage–Hybrid Approach: Processes, Key Inputs, Key Steps, and Key Outputs

The number of Processes, Key Inputs, Key Steps, and Key Outputs for the Check/Act Stage–Hybrid Approach should be determined by the Project Manager and approved by the Project Stakeholders (especially the Requesting Organization and the Project Sponsor) based on the relative Scope and Priority levels of the project.

SUMMARY

In this chapter, we covered the Check/Act Stage, its purpose and scalable framework, its 16 Processes, its nine Key Inputs, its seven Key Steps, and its five Key Outputs for both the "Full" and "Lean" Approaches.

We focused on the Check/Act Stage processes needed to check on the latest performance data for each agreed-upon periodic interval (weekly, bi-weekly, monthly, etc.) to determine the project's performance to date and forecast the expected values for being able to fully meet the project requirements in a timely and cost-effective manner, or not.

This Stage is actually a combination of BOTH "Check" AND "Act" because of the iterative nature of the conditional interactions between three different Stages: (1) between "Check" and "Execute"; (2) between "Check" and "Plan"; and, (3) between "Check" and "Confirm." Therefore, each one of those conditional interactions would be "acting on" the latest performance data.

Now, we close this chapter by relating how its contents apply to one of the two Case Studies: **Healthcare DCC.**

PROJECT CASE STUDY EXAMPLE: HEALTHCARE DCC

Here we resume describing how one of the co-authors addressed the challenges and applied the principles, practices, and processes for the Check/Act Stage to the Healthcare DCC Program comprised of a set of 59 projects divided up into three phases (see Figures I.3 and I.4), which were described in more detail at the end of Chapter 3 (see Figures 3.13 through 3.17) and again at the end of Chapter 6 (see Figure 6.5).

Check and Act On the Performance Data

As was mentioned at the end of Chapter 3, we met each week for 12 weeks during Phase I of the OneDCC Going Forward Plan Initiative and, during each weekly meeting, we used a color-coded tracking Gantt chart to:

1. Check with the Project Manager on the progress of those projects that were supposed to have "Started", be "In-Progress," or be "Completed" that *same* week.
2. Check with the Project Manager on the latest status of those projects that were supposed to be "Starting" or "Completed" the *following* week. (See Figure 8.5 for a shaded-grey version of a sample color-coded tracking Gantt chart as of the end of the third week of November of Year #5).

The "color-coded" set of progress/status symbols were based on the classic "traffic signal approach": "Green" = on or ahead of schedule; "Yellow" = in jeopardy of becoming behind schedule; and "Red" = behind schedule or missed a completion date.

Since we wanted to be sure we were NOT managing these projects by "looking at the review mirror" as can happen with the use of the "traffic signal approach," we kept the focus on each project's schedule performance as trends over time, not just a single snapshot. Since everyone had a say in finalizing these project schedules, the funding source was checking our progress regularly, and we had created a realistic sense of urgency, things were going more smoothly than we had anticipated.

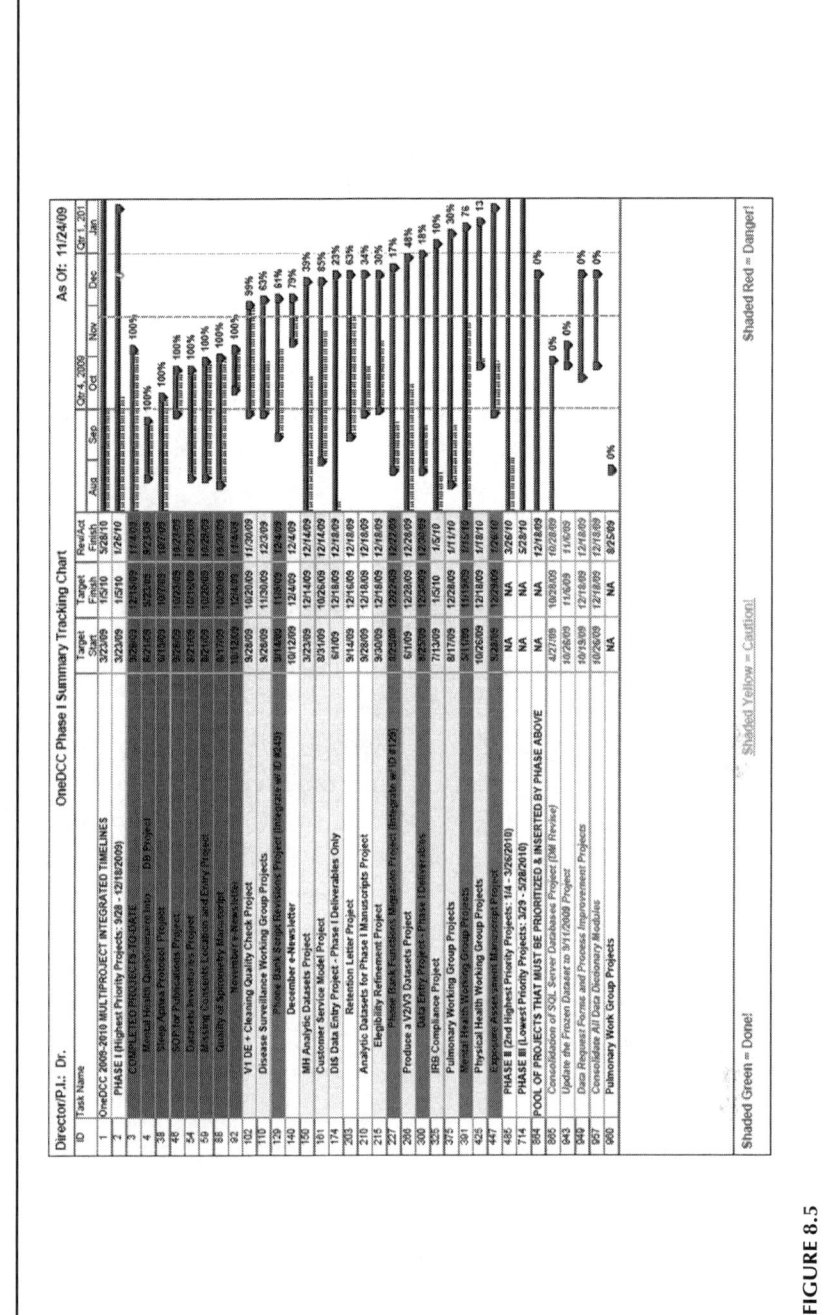

FIGURE 8.5
OneDCC Going Forward Plan Phase I color-coded Check/Act.

Believe it or not, this simple approach was sufficient to get the attention of each of the DCC Project Managers and Team Leaders. The preexisting "risk-averse," "playing-it-safe," "job-security-above-all" mentality mentioned in an early part of this Case Study actually worked in our favor at that point. Why? Because no one wanted to "suffer the embarrassment" of having their project (and, therefore, themselves) highlighted in either "Yellow" or "Red" in front of the Acting P.I. and the Core Leaders at one of the weekly meetings, without a legitimate reason or explanation. That might jeopardize their future "job security"! However, in those cases where there was an unforeseen schedule delay, we were able to persuade everyone to address it as OUR problem for which WE should help the Project Manager and his/her Team find and implement a solution together. It seemed to work!

While we did NOT Check on each project's budget performance during these Phase I weekly meetings, we DID Check and Act On their Scope, Schedule, and Quality performances directly, with an emphasis on Schedule due to the direct impact of project completions on our continued funding. Above all and in the long run our continued funding would be the ONLY budget or financial performance that meant anything to the One DCC Team!

Well, much to our surprise (but pleasantly so), we successfully completed 12 of the 17 highest-priority or most time-sensitive projects for Phase I before going on that pre-scheduled two-week holiday break. The other five had slipped into January 2010 and were marked as "DONE" and Phase I as "CLOSED" at our weekly status meeting on Friday, January 22. That Phase I portfolio of project performance FAR exceeded our expectations!

9

Stage #5: Confirm the Results (Iterate?)

INTRODUCTION

In this chapter, we focus on the processes needed to confirm the results to date and that:

1. The benefits that were promised in the Business Case and Benefits Management Plan are ready to be delivered
2. The opportunity for improvement (OFI) has been exploited, its goal has been reached, or the original problem has been solved
3. The Project Team has finalized all work activities and are ready to formally close the performance improvement project or the current Phase of the project.

If, upon completion of this Stage, there are remaining work activities for another Phase, you should iterate back up to the Align Stage as described in Chapter 5 and continue as illustrated in Figure 9.1.

As with the Align, Plan, Execute, and Check/Act Stages already covered, we have identified three scalable Approaches—*Full, Lean* or *Hybrid*—which can optionally be applied to this Stage depending on the project's relative Scope and Priority levels as per Figure 5.3. These options are documented below.

We close this chapter by relating how its contents apply to **both** of the two Case Studies introduced previously.

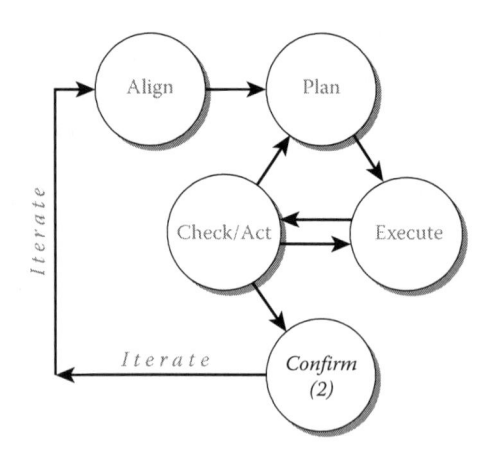

FIGURE 9.1
The Confirm Stage within APECC.

CONFIRM THE RESULTS

The primary purpose of this Stage is to confirm that the promised benefits have been delivered or are on pace to be delivered before closing the project or the current Phase of the project. This Stage has two Processes as illustrated in Figure 9.2.

Confirm Stage–Full: Processes, Key Inputs, Key Steps, and Key Outputs

The Confirm Stage only has two Processes, as depicted in Figure 9.2, making it the smallest of the five Stages. Yet, it is still an important one. If your organization has a Project Portfolio System (PPS) and/or a Portfolio Management Office (PMO), then it's likely that a variety of forms and templates will be available for you to use as "Key Inputs" to go through the "Key Steps" to produce the "Key Outputs" identified below. If you don't have one, then you'll either have to create them from scratch or contact one of us using the information provided at the end of the Preface of this book.

Once the objectives for change have been attained and reported, it's time to close out the project or the current phase. This last step is very often overlooked as a bit of change exhaustion may have set in during the course of the initiative. Understanding that this dynamic will be present as your

Stages / PM Domains	Align (3)	Plan (27)	Execute (13)	Check/Act (16)	Confirm (2)
1—Integration	Develop Project Charter	Develop Project Mgmt Plan	Manage Project Work Manage Project Knowledge	Monitor and Control Project Work Perform Integrated Change Control	Close Project/ Phase
2—Scope		Plan Scope Management Collect Requirements Define Scope Create WBS		Validate Scope Control Scope	
3—Schedule		Plan Schedule Mgmt Define Activities Sequence Activities Estimate Activity Durations Develop Schedule		Control Schedule	
4—Cost		Plan Cost Management Estimate Costs Determine Budget		Control Costs	
5—Quality		Plan Quality Management	Manage Quality	Control Quality	
6—Resources		Plan Resource Mgmt Estimate Resources	Acq. Resources Develop Team Manage Team	Control Resources	
7—Communications		Plan Communications Management	Manage Communications	Monitor Communications	
8—Risk		Plan Risk Management Identify Risks Perform Qualitative Risk Analysis Perform Quantitative Risk Analysis Plan Risk Responses	Implement Risk Responses	Monitor Risks	
9—Procurement		Plan Procurement Management	Conduct Procurements	Control Procurements	
10—Stakeholders	Identify Stakeholders	Plan Stakeholder Engagement	Manage Stakeholder Engagement	Monitor Stakeholder Engagement	
11—*Change*	Align for Change	Enroll For Change Plan Change Management	Manage Change	Check Change Management	*Confirm Change*
12—Technology		Plan Technology Management	Implement Technology Document Technology	Train Technology Report Tech Integrate Tech	

FIGURE 9.2
The Confirm Stage and its two processes.

project winds down and energy naturally shifts elsewhere, be sure to plan up front on documenting the lessons learned at the end of this Stage (transition or closure). This type of information is a critical part of the knowledge management system and the **Manage Project Knowledge** process in Project Integration Management.

Here are the nine Key Inputs, two Key Steps, and six Key Outputs for the Confirm Stage–Full Approach:

Key Inputs for the Confirm Stage–Full Approach

Before you can expect to effectively and efficiently confirm the results of an improvement project that meets the "Full" Approach criteria, you'll need to have the following nine Key Inputs that should have been either used or created during the Align Stage or the Check/Act Stage (for background, see Chapters 5 and 8, respectively):

- Project Charter, Accepted Deliverables, Agreements, Project Management Plan, Change Agenda, Project Business Case, Benefits Management Plan, Project and Procurement Documents, and Organizational Process Assets.

Key Iterative Steps for the Confirm Stage–Full Approach

We suggest the following simple sequence of two steps for completing this Stage in an orderly (and, if necessary, iterative) fashion as per Figure 9.3. The two Key Steps for the Confirm Stage–Full Approach are:

Step #1 (Change Management): Use the Change Management Plan, the Sustain-the-Change Management Plan, the Project Business Case, the Benefits Management Plan, and the Change Agenda as the Key Inputs to perform the **Confirm Change** process.

Step #2 (Integration Management): Next, you can use the Project Charter, Agreements, Project and Procurement Documents, and the Configuration Management Plan to perform the **Close Project/Phase** process to produce the six Key Outputs. (Iterate, if needed.)

Key Confirm Stage–Full Approach Outputs

The six Key Outputs from this Stage's "Full" Approach are the following:

- Change Alignment Confirmation Report, Sustain-the-Change Management Plan, Implementation Plan, Acceptance of the Final Improved Deliverable(s), Project Document Updates, and Organizational Process Asset Updates.

- PC = Project Charter
- PMP = Project Management Plan
- PPDs = Project and Procurement Documents
- ADs = Accepted Deliverables
- CA = Change Agenda
- BMP = Benefits Mgmt Plan
- As = Agreements
- PBC = Project Business Case
- OPAs = Org'l Process Assets

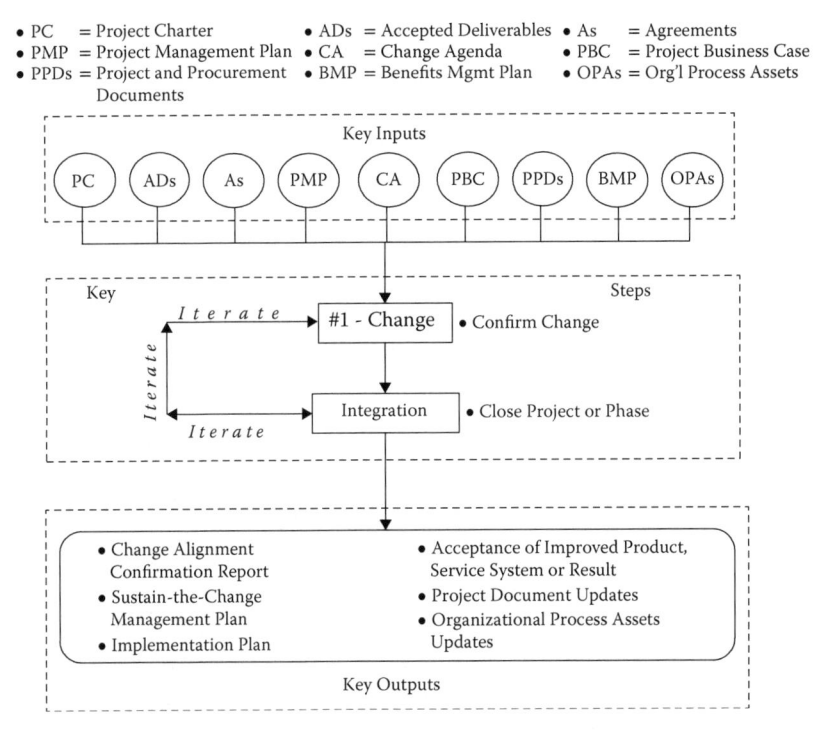

FIGURE 9.3
The Key Steps for the Confirm Stage–Full Approach.

Confirm Stage–Lean: Processes, Key Inputs, Key Steps, and Key Outputs

Here are the eight Key Inputs, one Key Step, and three Key Outputs for the Confirm Stage–Lean approach as per Figure 9.4.

Key Inputs for the Confirm Stage–Lean Approach

Before you can expect to effectively and efficiently confirm the results of an improvement project that meets the "Full" Approach criteria, you'll need to have the following eight Key Inputs which should have been either used or created during the Align Stage or the Check/Act Stage (for background, see Chapters 5 and 8, respectively):

- Project Charter, Accepted Deliverables, Agreements, Project Management Plan, Project Business Case, Benefits Management Plan, Project and Procurement Documents, and Organizational Process Assets.

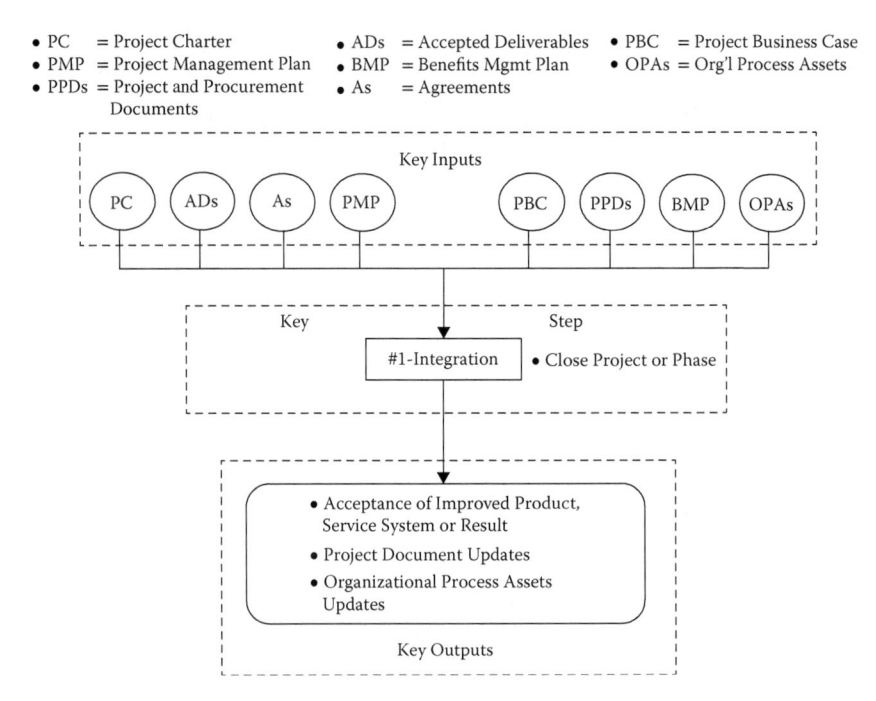

- PC   = Project Charter
- PMP = Project Management Plan
- PPDs = Project and Procurement Documents
- ADs = Accepted Deliverables
- BMP = Benefits Mgmt Plan
- As   = Agreements
- PBC = Project Business Case
- OPAs = Org'l Process Assets

FIGURE 9.4
The Key Steps for the Confirm Stage–Lean Approach.

Key Iterative Steps for the Confirm Stage–Full Approach

The one Key Step for the Confirm Stage–Lean Approach, as depicted in Figure 9.4, is:

Step #1 (Integration Management): You can use the Project Charter, Agreements, Project and Procurement Documents, and the Configuration Management Plan to perform the **Close Project/ Phase** process to produce the three Key Outputs. (Iterate, if needed.)

Key Confirm Stage Outputs–Lean Approach

The three Key Outputs from this Stage's Lean Approach are the following:

- Acceptance of the Final Improved Deliverable(s), Project Document Updates, and Organizational Process Asset Updates.

Confirm Stage–Hybrid Approach: Processes, Key Inputs, Key Steps, and Key Outputs

The number of processes, Key Inputs, Key Steps, and Key Outputs for the Confirm Stage–Hybrid approach should be determined by the Project Manager and approved by the Project Stakeholders (especially the Requesting Organization and the Project Sponsor) based on the relative Scope and Priority levels of the project.

SUMMARY

In this chapter, we covered the **Confirm Stage**, its purpose and its scalable framework, its two processes, its nine Key Inputs, its two Key Steps, and its six Key Outputs for both the "Full" and "Lean" approaches. We focused on the processes needed to confirm the results to date and whether or not the following statements are true:

1. The benefits that were promised in the Business Case and Benefits Management Plan are ready to be delivered.
2. The opportunity for improvement (OFI) has been exploited, its goal has been reached, or the original problem has been solved.
3. The Project Team has finalized all work activities and is ready to formally close the performance improvement project or the current Phase of the project.
4. The project has been terminated before completion because there was a high probability that the project would not produce the desired results.

If, upon completion of this Stage, there are remaining work activities for another Phase, you should iterate back up to the Align Stage.

PROJECT CASE STUDY EXAMPLES: HEALTHCARE DCC AND PHARMACUETICAL EPROVIDE II

Here we conclude describing how one of the co-authors addressed the challenges and applied the principles, practices, and processes for the Confirm Stage for *both* Case Studies:

- The healthcare MM&TP Data and Coordination Center portfolio of projects was described at the end of this book's Introduction (see Figures I.3 and I.4 as well) and and also discussed at the end of Chapters 3, 6, and 8.
- The pharmaceutical eProvide II Project was described at the end of this book's Introduction (see Figures I.5 and I.6 as well) and at the end of Chapters 4, 5 and 7.

Healthcare DCC: Confirm the Results

As already mentioned in this book's Introduction and Chapters 3, 6 and 8, for 12 weeks (September 28, 2009–December 18, 2009), we worked on executing 17 of the highest-priority or most time-sensitive projects identified in Phase I of the "OneDCC Going Forward Plan Strategic Initiative." We had intentionally scheduled a two-week holiday break afterwards so that everyone could take a "breather" before commencing Phase II following the New Year's celebration (see Figure 3.14 for the target dates for the three Phases, Figure 3.15 for multiproject rolled-up timelines for all three Phases, and Figure 3.16 for the Phase I multiproject integrated rolled-up timelines).

Well, much to our surprise (but pleasantly so), we successfully completed 12 of the 17 highest-priority or most time-sensitive projects for Phase I *before* going on that pre-scheduled two-week holiday break. The other five Phase I projects had slipped into January 2010 but they, too, were marked as "DONE" and the Phase I results were officially registered as "CONFIRMED" with the Acting P.I. and the Core Leaders at our weekly status meeting for the following projects on **Friday, January 22, 2010:**

1. Mental Health Questionnaire into ClinicalDB Project
2. Sleep Apnea Protocol Project
3. SOP for Publications Project

4. Manuscript Authors Datasets Inventory Project
5. Disease Surveillance Datasets Inventory Project
6. Missing Consents Location and Entry Project
7. Quality of Spirometry Manuscript Project
8. HP November e-Newsletter Project
9. Visit 1 Data Entry and Data Cleaning Quality Check Project
10. HP December e-Newsletter Project
11. Customer Service Model Project
12. Mental Health Analytic Datasets Project
13. DIS Data Entry for Rushless/Flushitt/Niceview Project
14. Visit 2/Visit 3 Datasets Production Project
15. Phase I Manuscripts Analytic Datasets Project
16. Portable Spirometry Calibration Manuscript Project
17. HP Retention Letter Project

The performance of the Phase I portfolio of projects FAR exceeded our expectations! (See a Project Team Photo in Figure 9.5.)

FIGURE 9.5
A Healthcare DCC Project Team.

Pharma eProvide II: Confirm the Results

As already mentioned in this book's Introduction and Chapters 4, 5, and 7, eProvide II was a "Top Ten", enterprise-wide, user provisioning and identity management project performed by the IT Global Shared

Services (GSS) Division of a Fortune 125 biopharmaceutical company. At the time our co-author was recruited to manage this project, user-provisioning implementations in small, medium, and large organizations were increasing due to Sarbanes-Oxley regulatory compliance needs and enhancements in identity protection and security, role management, reporting, and industry support.

It was selected as one of the "Top Ten" projects back in 2005–2006 based on its fulfillment of all seven required criteria that focused on *maximizing business value*:

1. Required cross-GSS coordination.
2. Had a major business impact on one or more clients.
3. Required the provision of capital assets.
4. Involved an investment greater than US$1 million.
5. Required multiple levels of governance.
6. Was a target solution within the IMprove strategic initiative.
7. Used the Project Management Framework (full version).

From a schedule perspective, the eProvide II project had three distinct "Go-Live" dates targeted...one for each of the three Key Deliverables:

May 2006: Changes to the R&DI Account Request Form (ARF) backend completed
July 2006: Changes to the R&DI Account Request Form (ARF) frontend completed
Dec 2006: Migration of all accounts from eProvide to eProvide II.

Few of us were surprised by the sweeping success of the enterprise-wide, eProvide II mega-project when, one-by-one, each of the above three "Go-Live" dates were met in 2006. Why not? Because we had EVERY ONE of the following Critical Success Factors (CSFs) for contemporary project management going for us AND, most importantly, *we didn't take them for granted*:

1. Senior Leadership Sponsorship and Top-Down Strategic Energy
2. Engaged Stakeholders and End-Users

3. Clear Business Requirements and Technical Specifications
4. High Levels of Emotional Maturity among the Project Team Members
5. A Mindset Predisposed Toward Problem-Solving, Improvement, and Optimization
6. An Agile and Resilient Workforce
7. Project Management Expertise and Know-How
8. Skilled Technical Resources
9. Proactive, Integrated, and Preventive Approach
10. The Right Set of Tools, Techniques, and Infrastructure

As a result of the confirmation of the results—Scope, Quality, Time, and Cost—presented in May, July, and December 2006, the eProvide II project was closed shortly thereafter. The GSS holiday parties that were held that year were extraordinarily festive affairs, especially in the buildings where the eProvide Project Team members worked.

What could have been an agonizing lesson in Murphy's Law (i.e., "Anything that CAN go wrong, WILL go wrong, and at the WORST possible moment") or even O'Malley's Corollary (i.e., "Murphy was an optimist!") on such a high-technology initiative turned out to be one of those rare, celebrated undertakings that don't come around often enough in one's career. That's why Figure 9.6 has so many smiling faces!

FIGURE 9.6
eProvide II Project Team.

10

Sustaining the Gains and Realizing the Benefits

INTRODUCTION

Now that you've learned about our "APECC" contemporary framework for managing performance improvement projects and programs in an iterative and scalable fashion to achieve successful outcomes, you may be wondering how you can ensure or sustain the gains that have been accomplished and the benefits that have been realized. Well, you may have forgotten about it but there's a crucial, pre-project document that was a Key Input both to the Align Stage back in Chapter 5 and the Confirm Stage back in Chapter 9 about which you may have forgotten: it's the **Project Benefits Management Plan**.

PROJECT BENEFITS MANAGEMENT PLAN[1]

A Project Benefits Management Plan describes how and when the project's business value, return on investment, or benefits will be delivered and how that business value will be measured. It may include one or more of the following sections:

- *Target Benefits*: The expected tangible and intangible business value to be gained by the implementation of the product, service, system or result
- *Strategic Alignment*: How the project benefits will support and align with the business strategies of the organization.
- *Timeframe for Realizing Benefits*: The benefits by Phase, short-term, long-term, and ongoing.

- *Benefits Owner*: The accountable person to monitor, record, and report realized benefits throughout the timeframe established in the Project Benefits Management Plan.
- *Metrics*: The measures to be used to show benefits realized, direct measures, and indirect measures.
- *Assumptions*: The factors expected to be in place or to be in evidence.
- *Risks*: The threats that may prevent the realization of benefits.

According to the Project Management Institute, an organization can deliver continuous value from outputs and outcomes once they transition back to the business by way of the following nine "good practices"[2] in Benefits Realization Management as per Figure 10.1:

1. Plan for the operational, financial, and behavioral changes necessary by project and program recipients to continue monitoring benefit performance.
2. Implement the required change control based on the defined level of tolerance, and take corrective action.
3. Perform a benefits assessment, which should include formally verifying that the benefits have been delivered and are being realized.
4. Facilitate continual improvement through ongoing knowledge sharing/knowledge transfer, including lessons learned.

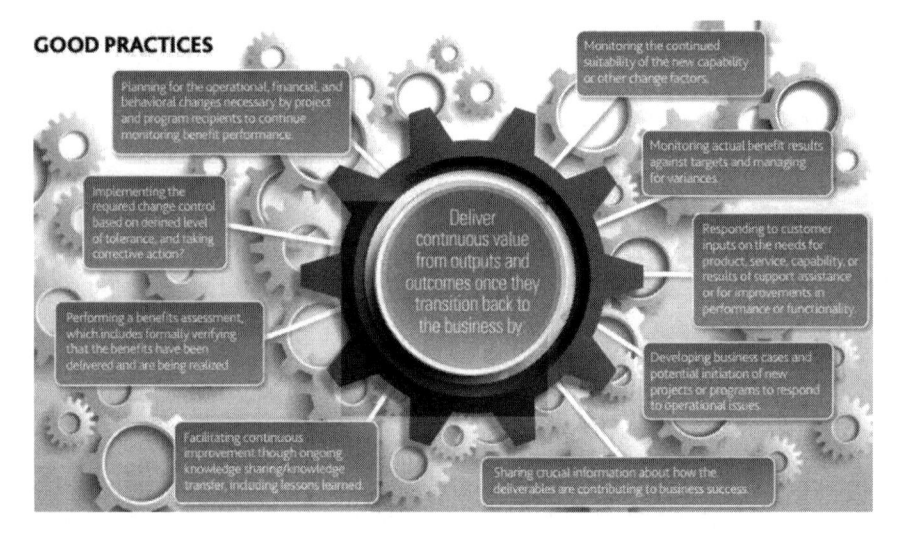

FIGURE 10.1

Sustain the gains via good practices in Benefits Realization Management.

5. Share crucial information about how the deliverables are contributing to business success.
6. Develop Business Cases and potential initiation of new projects or programs to respond to operational issues.
7. Respond to customer inputs on product, service, capability, or support assistance needs or concerning improvement in performance or functionality.
8. Monitor actual benefit results against targets and manage for variances.
9. Monitor the continued suitability of the new capability or other change factors.

SUSTAIN-THE-CHANGE MANAGEMENT PLAN

Another Key Input is the **Sustain-the-Change Management Plan,** which is a Key Output from the Confirm Stage in the contemporary PM4PITs framework (see Chapter 9). It provides structure to guide the post-project Implementation Plan as per the Benefits Management Plan, describing how to sustain the gains achieved during the performance improvement project that was approved by the Steering Committee.

IMPLEMENTATION PLAN

Finally, the **Implementation Plan** is another Key Output of the Confirm Stage in the contemporary PM4PITs framework (see Chapter 9) that will help the Project Team ensure that the gains achieved during the performance improvement project that was approved by the Steering Committee will be implemented as per the Benefits Management Plan and the Sustain-the-Change Management Plan.

SUMMARY

The **Project Benefits Management Plan** is a pre-project business document that is used in conjunction with the **Business Case** as a Key Input to the Align Stage and the Confirm Stage. It should be referenced along with the **Sustain-the-Change Management Plan** and the **Implementation Plan** after the project has been completed to ensure that the gains are sustained and the benefits realized going forward.

REFERENCES

1. Project Management Institute, *The Project Management Body of Knowledge* (aka *PMBOK® Guide*), 6th Edition, pp. 29, 33–34, 77, 125, 251, 469, 546, © 2017, Newtown Square, PA: Project Management Institute.
2. Project Management Institute, PMI Thought Leadership Series: Guiding the PMO, *Benefits Realization Management Framework*, © 2016, p. 5.

11

Connecting with the Organization's PMO

INTRODUCTION

One of the biggest wastes occurring today in organizations in both the private and public sectors is *the high percentage of failed projects and programs* with many of them *performance improvement projects and programs.* Billions of dollars are flushed down the drain and millions of employee hours are wasted each year that should be directed toward improving our work culture, our natural environment, and advancements in science, technology, engineering, and math (STEM). Redeploying 25 percent of these wasted resources to medical research, for example, could result in huge gains in decreasing mortality rates and extending life expectancy.

As we declared in a previous book,[1] there are many reasons why so many of these projects and programs have failed to deliver the desired benefits. Some of them include:

- Selecting the wrong projects and programs
- Defining the wrong opportunity for improvement or problem
- Not defining the best solution
- Lack of proper senior management attention
- Lack of an engaged sponsor
- Resource constraints and overallocations
- "Scope creep"
- Poor project management
- Lack of coordination between projects
- Not staffing projects and programs with the right personnel
- Inadequate funding

- Late projects that cause the organization to miss a crucial window of opportunity
- Poor organizational change management
- Inadequate training
- Poor use of technology
- Setting the wrong priorities
- Poorly defined goals and objectives
- Lack of understanding customer needs

This is just a partial list of things that can cause projects and programs to fail and we're quite sure we could go on extending the list, but you should have the idea by now. Three of the biggest unresolved problems that organizations face are **reducing cycle time**, **reducing costs**, and **reducing failure rates**.

The purpose of the preceding chapters of this book on PM4PITs has been to lay out how Project and Program Managers and their performance improvement teams can "*do their projects the right way, **one** project at a time*" and we hope we have been successful in doing so.

However, as we put the finishing touches on this book, we want to remind the reader one last time that you also need to be sure your organization is "*doing the right projects.*" This requires a continuous focus on optimizing the resources consumed by the organization's portfolio that is usually overseen by a Portfolio Development Team and a Portfolio Management Office (aka "Project Management Office") or PMO (see Figure 11.1).

Once an organization increases the emphasis on driving innovation, more performance improvement projects and programs are introduced into its system with increased pressure to implement them in shorter periods of time. With this increased emphasis on efficient and effective project management activities, the concept of a Portfolio Management Office and the use of Organizational Portfolio Management (OPM) has become a key element in an organization's success.

Well-managed, progressive organizations functioning in today's global work environment have realized that innovation is the key to a successful future. They have initiated communication and training systems to prepare their staff to be highly creative individuals who are prepared to take risks to make the organizations more successful. Organizaations are flooded with creative and innovative ideas from marketing, sales, product engineering, finance, information technology, product engineering, manufacturing engineering, and research and development.

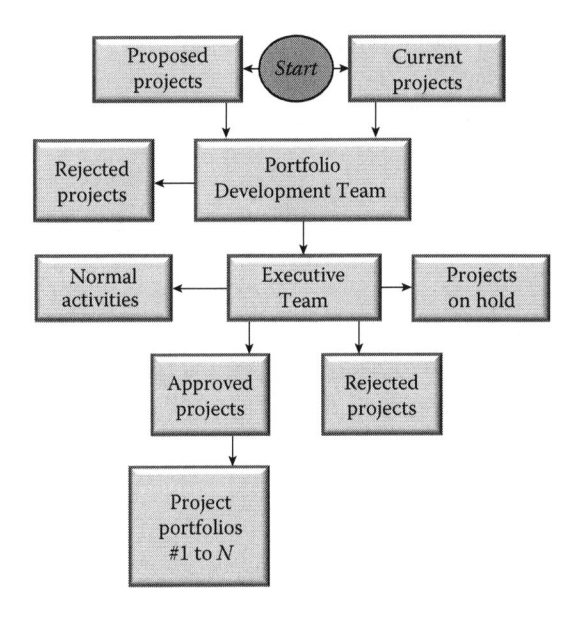

FIGURE 11.1
Overview of Organizational Portfolio Management.[2]

Each of these functional groups has had a Value Proposition prepared for one or more proposed projects that screened out and rejected the ideas that did not provide acceptable value-added content to the organization. The approved projects have had Business Plans prepared for them by an independent group that indicates that they have a high probability of adding value to the organization. Many of them are short-term, easy-to-implement concepts that are refinements to current projects. Others will require a long-term investment with the known risks associated with them. The ones that were successful in completing the Business Plan review should all be aligned with the organization's strategic plan, especially its mission, vision, and values (MVV).

The problem at that point rests with selecting the *right* projects and programs that will provide the maximum value to the organization—both short- and long-term—based upon the resources that can be made available to support those projects and programs. While those that successfully completed the Business Plan Review process have the potential of providing an acceptable added-value level to the organization, they cannot all be approved simultaneously due to the limited availability of funds, staffing, and facilities. These become the "projects on hold." As a result, the organization needs to select an Organizational Portfolio that can be governed by a Portfolio Management Office (PMO) that maximizes the values that can be created from the implementation of the projects and programs that make up the portfolio.

We urge you to stay connected with your organization's PMO to be sure you're "*doing the right projects the right way, **one** portfolio and **one** project at a time*" and we hope you'll be successful in doing so.

REFERENCES

1. Christopher F. Voehl, H. James Harrington, and William S. Ruggles, *Effective Portfolio Management Systems*, Preface, pp. xi–xii, © 2016, New York, NY: Productivity Press.
2. Ibid., p. 7.

Epilogue

We have now come to the end of the "Authoring Phase" of the latest addition to the "Little Big Book Series" (inaugurated by co-author H. James Harrington several years ago): *Project Management for Performance Improvement Teams* (PM4PITs). The manuscript files—both text and figures—were uploaded to the designated Dropbox and passed through the hands of the Executive Editor's assistant. As they've done so many times before, the publisher's editorial staff worked their magic to transform our raw, nonfiction wordage into a refined, practical, softcover book.

30TH ANNIVERSARY CELEBRATION

Writing this manuscript has been an intense labor of love, especially for me (co-author Bill Ruggles). That's because, only a few weeks ago, I marked the 30th anniversary of having received Certification as a Project Management Professional (PMP® #133). My original hope was to have the finished PM4PITs book hit the streets about this time but I persuaded our superb Executive Editor, Michael Sinocchi, that it would be better, in the long run, if it were to reference the content of PMI's latest release of *The Guide to the Project Management Body of Knowledge* (aka *PMBOK® Guide*, 6th Edition).

That meant waiting until that standard-setting guide became available in early September 2017 in conjunction with the release of the first edition of the *Agile Practice Guide*, co-published by PMI and the Agile Alliance. I'm convinced it was worth the wait and the resulting delay in releasing the book you are now holding in your hands. So, I invite you to join me in celebrating *the year after* my 30th anniversary as a Certified Project Management Professional.

HEALTHCARE PROGRAM DCC CASE STUDY

The Healthcare Program Data and Coordination Center Case Study concluded with the completion of Phase I—a portfolio of the 17 highest-priority projects in January 2010—as reported at the end of Chapter 9. But you may be wondering, "What happened to the Healthcare Program's DCC and its Phase II and Phase III projects?"

Well, the very *next week*, in late January 2010, the Acting Principal Investigator made an announcement at the weekly DCC staff meeting that altered the risk-accepting, deliverable-focused, results-driven mind-set and work culture we had developed the previous year: the federal funding source had notified him that we would NOT have to compete for renewal of the grant for the next funding year that would begin in July 2010. In sum: there was NO longer a sense of urgency and a compelling need to produce results in a timely fashion. The incumbent DCC staff breathed a collective sigh of relief, confident that they could go now back to their risk-averse, job security–focused, relationships-driven mind-set and the concomitant work culture they had enjoyed previously.

Hence, as you can surmise, we had to revise our Change Agenda, Enrollment Plan, Change Management Plan, and other related Plan Stage documents for Phases II and III and it became increasingly difficult to motivate the rank-and-file staff to get those 42 remaining projects and programs done in a timely fashion. Yet, with a lot of diplomacy and finesse (and even some begging), we were able to complete about 80% of them by the end of calendar year 2010.

Coinciding with the end of that calendar year, the outgoing U.S. Congress finally passed binding legislation that would ensure long-term funding (for what had been an ad-hoc Healthcare Program until then) that would commence with the start of the next funding year in July 2011. Unfortunately, at least for me, funding for a Project or Program Manager position was NOT included in the "write-up" of the bill and I began to seriously look at my career options in early 2011. The only opportunity that DID open up within the Healthcare Program, itself, was a Quality Assurance Manager vacancy. but I knew that I would not be the preferred choice for the person in charge of managing that group. So, I left the DCC in November 2011.

That Healthcare Program Consortium continues to function to this day and now services over 45,000 sickened patients.

PHARMACEUTICAL ePROVIDE II CASE STUDY

The second Case Study also concluded at the end of Chapter 9 with our declaration that it was "one of those rare, celebrated undertakings that don't come around often enough in one's career." Yes, it's TRUE!

If Murphy's Law and O'Malley's Corollary (see the last paragraph of Chapter 9 if you aren't familiar with them) have become all too common in 21st-century high-technology projects and programs, this one was characterized by what I call the "P. I. P. Approach": "If you approach what looks like an overwhelmingly challenging project with a Proactive, Integrated, and Preventive mind-set and you have the top-down support of your organization's Senior Leadership Team, you SHALL overcome the struggles and challenges that will present themselves along the way."

You just have to be vigilant and remember that, until you've confirmed that the promised benefits have been delivered and verified, your project is only as successful as it was YESTERDAY. Don't get complacent until the project is truly DONE!

There are two other relevant details we left out of the end of this Case Study in Chapter 9 that we think you'll find intriguing and, perhaps, amusing.

First, our co-author was NOT able to see the eProvide II project through to its official end in December 2006. Why not? Due to "funding constraints"! As a contracted Subject Matter Expert, there was a limited budget available for the use of his services and those funds ran out in August 2006, not long after the second of the three Key Deliverables had been met. As a result, he had to cheer on the eProvide II Project Team "from the sidelines" those last four months via periodic e-mails and phone calls with various Team Members.

Lastly, in July 2007—less than a year after he had to be let go due to the lack of sufficient funds—our co-author (who is also a Master Black Belt) was invited to be a member of a cadre of nine Black Belts and Master Black Belts on the "Lean Sigma Transformation Initiative Team" at the very same Fortune 125, pharmaceutical company.

On the first day of this latest consulting assignment for the same company, he was directed by his new boss to go through the official "on-boarding" process that he had arranged with the Procurement Department to obtain all of his credentials (e.g., ID badge, computer network access user ID and password, parking decal, office keys, etc.). When our co-author showed up to fulfill this requirement and told the Procurement Officer that he had previously served as a contractor, she checked the contractor database and informed him that, according to the record attached to his name, his prior credentials hadn't been turned in and closed out yet! He was told that unless he reconciled their current disposition, she could NOT **provision** his new credentials.

That word "provision" was the key to solving the mystery of the whereabouts of his missing credentials. He called his former boss—the Associate Director for *Provisioning Services*—and asked him if he knew what had happened to his credentials after he had turned them in 11 months earlier? Seated at his desk while still talking on the phone, his former boss turned to his right, opened the middle desk drawer, and VOILA…there they were, still sitting where he had placed them back in August of the previous year. He assured our co-author that he would access the eProvide II application to "off-board" him immediately and would turn in his ID badge and office keys that same day so that he could be "on-boarded" again.

Moral of this Case Study (and this book): addressing one opportunity for improvement (OFI) could lead to the creation of yet another! Both Continual and Continuous Improvement are NOT destinations; each of them is a JOURNEY!

CONCLUSION

We hope you have enjoyed and benefitted from this book and, as a result of deploying (or, even better, ADOPTING) its principles, processes, and practices, you will see a measurable improvement in the success of your performance improvement projects! Please let us know how your project and program teams perform using what we've shared in this book! You may reach us at:

- William S. ("Bill") Ruggles: bill.ruggles@ruggles2llc.com or (201) 956-7905
- H. James ("Jim") Harrington: hjh@svinet.com or (408) 358-2476

Glossary of Terms

The following terms and their definitions have been excerpted and expanded upon from the Project Management Institute's (PMI's) Guide to the Project Management Body of Knowledge (*PMBOK® Guide*)[1]:

Adaptive: A type of project in which the product is developed iteratively with detailed scope defined for each iteration as it begins with shorter intervals than the "predictive" approach (Also known as an "Agile" or "Scrum" approach).

Adaptive Life Cycle: A "stochastic" approach to project and portfolio management. For example, agile or change-driven methods respond to lots of change and stakeholder involvement. Adaptive is also interactive and incremental, but iterations have fixed costs and time (2–4 weeks), performing multiple processes in each iteration. Early iterations focus on planning.

- Scope is decomposed into requirements and WBS (see *Work Breakdown Structure*) to create the product backlog.
- Each iteration begins with prioritizing which high-priority items can be delivered and ends with a delivered product for the customer to review; it should have finished, complete, usable features.
- Sponsor and/or customer representative provides feedback on deliverables and status of backlog items.
- Preferred in rapidly changing environments when scope and requirements are difficult to define.

Advocate: The individual or group that wants to achieve a change, but lacks the power to sanction it. (See also *Change Agent* and *Project Champion*.)

Affected Individuals: These are the individuals whose activities will be directly affected by the output from the project/program when the project/program is completed. (See also *Impacted Individuals* and *Stakeholders*.)

Align Stage: A Stage in the contemporary Project Management Framework in which the strategic alignment of a new project or a new Phase of an existing project is ensured. This alignment is achieved by approving a Business Case that verifies the proposed project benefits and obtains authorization to start the project or continue to plan the next Stage of the project. It identifies the "opportunity for improvement," "goal," "pain point," or "problem" that is being addressed by the project; a Project Charter, a Stakeholder Register, and a Change Agenda should also be created during this Stage.

Benefits Management Plan: The documented explanation that defines the processes for creating, maximizing, and sustaining the benefits provided by a project or program. It formally documents the activities necessary for achieving the program's planned benefits. It identifies how and when benefits are expected to be delivered to the organization and specifies mechanisms that should be in place to ensure that the benefits are fully realized over time. It is the baseline document that guides the delivery of benefits during the program's performance.

Business Case: A documented evaluation (pre-project) of the potential impact a problem or an opportunity has on the organization to determine if it is worthwhile investing the resources to correct the problem or take advantage of the opportunity for improvement. It captures the reason for initiating a potential project or program. An example of the results of the Business Case analysis of a software upgrade could be that it improves the software's performance as stated in the Value Proposition, but it would (a) decrease overall customer satisfaction by an estimated 3%, (b) require 5% more task processing time, and (c) reduce system maintenance costs by only $800 a year. As a result, the Business Case does *not* recommend including the project in the portfolio of active programs. Often the Business Case is prepared by an independent group, thereby giving a fresh and unbiased analysis of the benefits and costs related to completing the project or program. (See *Value Proposition.*)

Business Objectives: Measureable targets used to define what the organization wishes to accomplish, often over the next 5 to 10 years.

Business Plan: A formal statement of a set of business goals, the reason they are believed to be obtainable, and the plan for reaching these goals. It also contains background information about the organization and/or services that the organization provides as viewed by the outside world.

Business Value: Entire value of the business; total sum of tangible (assets, fixtures, equity, utility) and intangible elements (goodwill, recognition, public benefit, trademarks): short, medium, or long term.

Cascading Sponsorship: The alignment of a project via the transformation or conversion of a change target into a sustaining sponsor (as depicted in Figure 3.7).

Change Agenda: A literal agenda of events that gets distributed as part of the Enrollment Plan to all prospective Change Agents (CAs). It sets the stage for meetings or conference calls held with CAs to explain the purpose, scope, and milestones of the project; engage them in a dialog to enroll them in the effort; explain their roles; and show them how they will be active participants in crafting the plan for change.

Change Agent: The individual or group serving as a catalyst to some type of change and responsible for facilitating the implementation of that change. (See also *Advocate, Project Champion.*)

Change Control Board: In information technology and software development, a Change Control Board (CCB) or Change Control Committee is an empowered group that makes decisions regarding whether or not proposed changes to a project should be implemented, evaluating potential impacts to stakeholder groups or other systems. In many cases, the CCB also reviews results of implemented changes, and evaluates corrective actions in the event planned changes have any adverse impacts.

Change Management Plan: A component of the overall Project Management Plan, it should define two perspectives describing how the Project Manager should (1) monitor and control changes TO the performance improvement project, especially its Scope, Schedule, and Budget performance baselines (addressed in Project Integration Management); and (2) monitor and control changes CAUSED BY the performance improvement project and their impacts on the Change Targets. It works in tandem with the Change Agenda and Enrollment Plan.

Change Target: See *Target.*

Check/Act Stage: A Stage in the contemporary Project Management Framework in which the processes are performed to review and analyze the results of the Execute Stage for the current periodic interval and identify what has been learned. Then, depending on the results for that periodic interval, take one of these three actions (Act On) to either (a) execute the work for the next periodic interval, (b) revise the Project Management Plan for the work in the next periodic interval prior to executing it, or (c) confirm that the work in the Project Management Plan has been completed.

Closing: A Process Group in the traditional Project Management Framework in which final project audits, project evaluations, product validations, acceptance criteria, and lessons learned are addressed. Even though it only contains ONE process, it is still referred to as a "Process Group".

Common Experiences: Shared mission, vision, and values (MVV) and beliefs; regulations, process/policy/procedure (PPP); common recognition and reward system; risk tolerance; view of leadership/authority; code of conduct, work ethic, hours; operating environments.

Communications Management Plan: A component of the Project Management Plan that describes how, when, by whom, and to whom information about the project will be administered and disseminated.

Competing Constraints: There are six of them comprised of the project's *scope, schedule, cost, resources, quality and risk.* They should be **balanced** or **tailored** by the Project Manager in collaboration with the Project Team, the Sponsor, or some combination thereof.

Confirm Stage: A Stage in the contemporary Project Management Framework in which the processes which verify that the benefits promised in the Business Case have been delivered, the project's opportunity for improvement has been exploited, its goal has been reached, or the original problem has been solved are performed. If there is another Phase, iterate back to the Align Stage and continue the iterative cycle.

Continual Improvement[1]: Improvements of various sizes achieved periodically by Project Teams; a traditional approach such as the "Plan–Do–Check–Act" (PDCA) cycle defined by Shewhart and the "Plan–Do–Study–Act" (PDSA) cycle defined by Deming used to identify and make changes that result in better, faster, less costly, or smarter project outcomes. It can be applied toward making continual, incremental enhancements in the quality of project management or the quality of a product, service, system, or result and assumes a duration that continues over a lengthy period of time but *with* intervals of interruption between projects.

Continuous Improvement: Little improvements achieved every day by everyone; a technique, such as Lean, Kaizen, 5S, Value Stream Mapping, etc., used as an ongoing effort to increase the efficiency of a process by eliminating waste and/or non-value-added activities. It assumes a perpetual work duration that continues *without* interruption.

Creativity: Idea generation. The use of the imagination or original ideas to design or produce something unique.

Critical Success Factors: The key things that the organization must do extremely well to overcome today's problems and the roadblocks to meeting the Mission and Vision Statements.

Cycle of Improvement: The iterative process of turning opportunities for improvement into strengths.

Development Approach: The method used to create and evolve the product, service, or result during the project life cycle, such as predictive (waterfall), iterative, incremental, agile, or a hybrid method. (See also *Project Life Cycle.*)

Due Diligence: The care that a reasonable person would exercise to avoid harm to other persons or their property.

Enrollment Plan: It details how the enrollment of stakeholders will be carried out in order to facilitate the change(s) being CAUSED BY the performance improvement project and its outcome(s) on the Change Targets. It works in tandem with the Change Agenda and Change Management Plan.

Enterprise Environmental Factor: A condition that is not under the immediate control of the project manager or the team but which may influence, constrain, or restrict the project.

Execute Stage: A Stage in the contemporary Project Management Framework in which the processes are performed to complete the work defined in the Project Management Plan in order to attain the project's objectives.

Executing: A Process Group in which the processes are performed to complete the work defined in the Project Management Plan in order to attain the project's objectives.

IaaS (Infrastructure-as-a-Service): A cloud-based service comprised of highly automated and scalable computer resources, complemented by cloud storage and network capability that can be self-provisioned, metered, and available on-demand.

Impacted Individuals: These are the individuals whose activities will be directly impacted by the output from the project/program when the project/program is completed. (See also *Affected Individuals* and *Stakeholders.*)

Implementation Plan: A Key Output of the Confirm Stage in the contemporary framework to ensure that the gains achieved during the performance improvement project that were approved by the Steering Committee are implemented as per the Sustain-the-Change Management Plan.

Initiating: A Process Group in the traditional Project Management Framework in which the processes are performed to define and authorize the start of a new project or a new Phase of an existing project.

Initiating Sponsor: The individual/group who has the power to initiate and legitimize the performance improvement project at the outset and the change(s) it will create for all the affected Change Targets. (See also *Project Sponsor* and *Sustaining Sponsor.*)

Innovation: Implementing ideas in ways that create economic value in your business world. The process by which people create value through the implementation of new and unique ideas. It is how an organization adds value to creative ideas. (*Note*: Innovation can take many forms: it can be a new idea, an insight, or a rearrangement of present ideas and/or hardware as long as it is new or unique, implemented, and creates significant value to the stakeholders and consumers. It applies to most activities including personal and organizational ones.

Iterative and Incremental: An agile process in which project phases/iterations repeat project activities as the team's knowledge increases and its understanding of the product improves. Iterations develop the product in repeated cycles and increments add to the functionality in succession. All Project Management Process Groups are performed.

- Deliverable(s) are produced at the end of each iteration, and each iteration incrementally builds the deliverables until the exit criteria for the Phase are met. The Project Team processes feedback at the end of each iteration.

- A high-level vision is prepared, but the detailed scope is elaborated one iteration at a time, planning for the next iteration as work is being done on the current one (managing and confining the scope).
- It is preferred for projects with changing scope and objectives, to reduce complexity, or when partial delivery is beneficial and does not impact the final outcome deliverable(s). Reduces risk on large, complex projects by incorporating lessons learned after each iteration.

Key Performance Driver (KPD): A functional factor within the organization that management can change that controls or influences the organizational culture and the way the organization performs.

Key Performance Indicator (KPI): A quantifiable measure used to evaluate the success of an organization, strategic initiative, employee, etc., in meeting the objectives for performance.

Knowledge Base: Including configuration management, versions, and baselines; financial information (hours, costs, budgets, overruns); lessons learned; issue/defect databases; process data; prior project files.

Manager: An individual who accomplishes an assigned task or set of tasks through the use of other individuals to whom the work is delegated.

Mission Statement: The stated reason for the existence of the organization. It is usually prepared by the CEO and key members of the executive team and succinctly states what they will achieve or accomplish. It typically is changed only when the organization decides to pursue a completely new market.

Monitoring/Controlling: A Process Group in the traditional Project Management Framework in which the processes are performed to track, review, and regulate the progress and performance of the project and identify any areas in which changes to the Project Management Plan are required, and initiate any related changes.

Ongoing Work: A repetitive process following existing procedures.

Operations: Ongoing work, production of goods and services. Generally out of scope from a project/program/portfolio management standpoint. Operational stakeholders should be added to the stakeholder register and their influence (pro or con) addressed as a risk.

Opportunity for Improvement (OFI): An advantageous opening, suggestion or, in some cases, a recommendation to help an organization enhance the current state of a particular process, system, product, or service such that it will perform better in a prospective future state. It is a concept that has been made popular by the Baldrige Performance Excellence Program.

Organization: A systematic arrangement of entities (people, departments, divisions, companies, etc.) aimed at accomplishing a common purpose, which may involve undertaking projects.

Organizational Goals: They document the desired, quantified, and measurable results that the organization wants to accomplish in a set period of time to support its business objectives (for example, increase sales at a minimum rate of 12 percent per year for the next 10 years with an overall average annual growth rate of 13 percent). Goals should be specific rather than general so that there is no ambiguity.

Organizational Master Plan: The combination and alignment of an organization's Business Plan, Strategic Business Plan, Combined Performance Acceleration Management (PAM) Plan, and Annual Operating Plan.

Organizational Planning: Impacts projects; prioritization based on risk, funding, and impact on strategic plan objectives.

Organizational Portfolio Management (OPM): The combined coordination and management of all the active projects and programs to maximize the value they add to the organization by continuously monitoring their progress, prioritizing work, and allocating resources. It refers to the combined activity of all the active portfolios and independent projects going on within the organization and the improvement of the organization's capability linking project/program/ portfolio management with organizational facilitators (structural, cultural, technological, human resources practices) to support strategic goals. To apply this methodology, organizations must first measure their current capabilities, identify the future target capabilities, and then plan and implement improvements to close the gap between the two. (See also *Project Portfolio Management (PPM)*)

Organizational Portfolio Management (OPM) System: An automated or electronic application that enables the centralized oversight of processes, methods, and technologies by Portfolio Leaders,

Program Managers, Project Managers, and Project Management Offices (PMOs) to concurrently analyze and manage all proposed and active projects. (See also *Project Portfolio Management (PPM) System*.)

Organizational Portfolio Manager: The individual that is assigned to manage the organization's portfolio and is held responsible for Organizational Portfolio Management activities. In organizations that have a Project Management Office, this individual is often referred to as the PMO Manager or Director. (See also *Project Portfolio Management (PPM) System*.)

Organizational Process Assets: Plans, policies, procedures and processes, and knowledge bases specific to or used by the performing organization. Any artifact, practice, or knowledge used on the project policies, procedures, and processes.

PaaS (Platform-as-a-Service): A cloud-based service that typically provides a platform on which software can be developed and deployed.

Performance Improvement Team: A special type of Project Team that includes the Project Manager and members of a team of people responsible for completing a performance improvement project successfully. Roles include Analysts, Leads, Back-Ups, SMEs, Contractors, Business Partners, etc.

Performance Measurement Baselines: An approved set of integrated plans for the project's Scope, Schedule, and Budget against which project execution will be compared to measure and manage performance. They include "contingency reserve" but not "management reserve" allocations.

Performing Organization: The entity that possesses the knowledge, skills, tools, and techniques required to submit a proposal, quote, or bid to execute the work involved in completing a project in accordance with the expectations or requirements provided by a Requesting Organization. Also known as the "Seller," "Supplier," "Provider," or "Vendor."

Plan Stage: A Stage in the contemporary Project Management Framework in which a detailed "roadmap" (Project Management Plan) to guide execution is prepared for exploiting the "opportunity for improvement," "goal," "pain-point," or "problem" that was identified in the Align Stage.

Planning: A Process Group in the traditional Project Management Framework in which processes are performed to define the project's Scope, Quality, Schedule, Budget, and Risk objectives and how best to attain them via the Project Management Plan.

Policy: A principle or rule to guide decisions and achieve rational outcomes. A policy is an intent to govern, and is implemented as a procedure. Policies are generally adopted by the Board of Directors or senior governance body within an organization, whereas procedures are developed and adopted by senior managers. Policies can assist in both subjective and objective decision making.

Portfolio: A centralized collection of independent projects or programs that are grouped together to facilitate their prioritization, effective management, and resource optimization in order to meet strategic organizational objectives. (See also *Project Portfolio.*)

Portfolio Components: Constituent programs, projects, and other related work. Status reports, lessons learned, and change requests roll up to the portfolio.

Portfolio Development Team: An executive-level group that reviews/prioritizes projects and programs for resource allocation that is aligned to organizational strategies.

Portfolio Leader: A senior Project or Program Manager qualified and appointed to manage multiple concurrent and interdependent sub-portfolios, programs, and projects. Portfolio Leader roles are typically awarded to Program Managers with years of demonstrated success organizing and managing programs with multimillion-dollar budgets allocated from (and aligned to) the organization's key Strategic Objectives.

Portfolio Management: Aligns organizational strategy by prioritizing programs and projects, prioritizing work, and allocating resources. It is the "centralized management of one or more portfolios to achieve strategic objectives."

Portfolio Management Office (PMO): A centralized management structure that standardizes project-related governance policies and procedures and facilitates the sharing of resources, methodologies, templates, tools, and techniques in one or more portfolios. There are three different types of Portfolio Management Offices:

- *Supportive* (low control; consulting; providing a P/T mentor who can provide advice, standardized templates, and maintain a project status repository)

- *Conforming* (moderate control; validating; ensuring compliance with SOPs and other project management–related standards)
- *Directive* (high control; authoritative; actively managing and leading projects)

Portfolio Manager: A senior Project Manager qualified and appointed to manage multiple concurrent and interdependent sub-portfolios, programs, and projects. Portfolio Manager roles are typically awarded to program managers with years of demonstrated success organizing and managing programs with multimillion-dollar budgets allocated from (and aligned to) the organization's key Strategic Objectives. (See also *Organizational Portfolio Manager.*)

Portfolio Steering Committee: An executive committee that is responsible for overseeing all the active projects/programs. All major changes in timing, funding, and an output from the project process needs to be approved by this committee.

Predictive: A type of project in which a plan-driven product and its deliverables can be defined at the beginning of the project and whose scope must be carefully managed until the end.

Predictive Life Cycle: A "deterministic" approach to project and portfolio management (for example, requirements, feasibility, planning, design, construct, test, turnover); preferred when the outcome is well understood, a base of industry practice exists, or the outcome needs to be delivered in full in order to have value to stakeholders ("waterfall" approach).

Process: A set of interrelated actions and activities performed to create a pre-specified product, service, or result. Each process is comprised of inputs, tools, and techniques (or activities), and outputs with constraints (environmental factors), guidance, and criteria (organizational process assets) taken into consideration.

- Select appropriate processes to meet project objectives, adapt a defined approach to meet requirements, communicate/engage stakeholders, meet needs, balance constraints.
- Project management processes: ensure flow throughout the project life cycle; this includes tools and techniques to apply PM skills and capabilities.

- Product-oriented processes: specify and create the product, defined by project life cycle; these vary by application area and project Phase. Required to define the scope but not defined in the *PMBOK® Guide*.

Program: A group of related projects, subsidiary programs, and program activities managed in a coordinated way to obtain benefits not available from managing them individually. May include work outside the scope of projects but will always have two or more projects within its scope.

Program Charter: The document, issued by the Program Sponsor, that formally authorizes the existence of a program and provides the Program Manager in the Performing Organization with the authority to apply organizational resources to program activities on behalf of the Requesting Organization. It establishes a partnership between the Performing Organization and the Requesting Organization.

Program Management: The application of knowledge, skills, and principles to a program to achieve the program objectives and to obtain benefits and control not available from managing them individually. Focuses on:

- Project interdependencies.
- Resolving resource constraints/conflicts among projects.
- Aligning strategic direction to impact PPP goals and objectives.
- Addressing issues and changes within the governance structure.

Program Manager: A person authorized by the Performing Organization to lead the team or teams responsible for achieving program objectives. Program Managers are typically responsible for organizing and managing the projects under a unifying program to best manage constrained resources across multiple projects. (See also *Program Management*.)

Project: A temporary endeavor undertaken to create or modify a unique product, service, system, or result.[1] Therefore, there is no such thing as an "ongoing" or "never-ending project"; that would be an "anti-project." Projects must have a defined beginning and end that drives some type of change. A project can create a product, service, improvement, or result (e.g., a plan or document).

Project Benefits Management Plan: A pre-project business document (along with the Business Case) that defines the processes for creating, maximizing, and sustaining the benefits provided by a project or program. It describes the expected plan for realizing the benefits in the Business Case and formally documents the activities necessary for achieving the project's planned benefits. It identifies how and when benefits are expected to be delivered to the organization and specifies mechanisms that should be in place to ensure that the benefits are fully realized over time. (See also *Benefits Management Plan.*)

Project Champion: An organizational leader with a demonstrated stake in the implementation and sustainability of a portfolio, program, project, or product. Champions will generally exert their influence to remove barriers to success and break down resistance to change. Ideally they will align with other leaders to ensure enrollment and adoption throughout the organization. (See also *Advocate* and *Change Agent.*)

Project Charter: The document, issued by the Project Sponsor, that formally authorizes the existence of a project and provides the Project Manager in the Performing Organization with the authority to apply organizational resources to project activities on behalf of the Requesting Organization. It establishes a partnership between the Performing Organization and the Requesting Organization.

Project Life Cycle: The series of generally sequential Phases a project passes through from beginning to end. Starting, organizing and preparing, performing project work, closing; cost and staffing levels low at the start and end; risk and uncertainty greatest at the start; ability to influence highest at start; later changes cost more. (See also *Development Approach.*)

Project Management: The application of knowledge, skills, tools, and techniques to balance competing constraints in order to meet project requirements and, when applicable, project portfolio priorities. These constraints include scope, quality, schedule, cost, resources and risk. PMI has identified a total of 49 unique project management processes within five Process Groups: Initiating, Planning, Executing, Monitoring/Controlling, and Closing.

Project Management Office (PMO): A project-oriented management structure that standardizes project-related governance policies and procedures and facilitates the sharing of resources, methodologies, templates, tools, and techniques. There are three different types:

- *Supportive* (low control; consulting; providing a P/T mentor who can provide advice, standardized templates, and maintain a project status repository)
- *Conforming* (moderate control; validating; ensuring compliance with SOPs and other project management-related standards)
- *Directive* (high control; authoritative; actively managing and leading projects)

Project Management Plan (PMP): This is the most important document produced by the Develop Project Management Plan process in the Planning Process Group or the Plan Stage. It contains the outputs from the other 26 planning processes. Due to the iterative nature of planning, it is progressively elaborated, improving and detailing this output as new information and estimates become available, defining and managing work at a detailed level until the three Performance Baselines (Scope, Time, and Cost) have been set. Then, it may only be changed when approved by the Change Control Board.

Project Management Processes: Ensure an iterative flow of work and data throughout the project, including the efficient use of PM knowledge, skills, tools, and techniques.

Project Management Process Group: A logical grouping of project management processes in the traditional framework, each of which includes one or more inputs, tools and techniques, and outputs, but they should *not* be confused with "project Phases." There are five of them in the traditional framework:

1. **Initiating:** Those processes performed to define and authorize the start of a new project or a new Phase of a project.
2. **Planning:** Those processes performed to define the project's Scope, Quality, Schedule, Budget, and Risk objectives and how to attain them via the Project Management Plan.
3. **Executing:** Those processes performed to complete the work defined in the Project Management Plan in order to attain the project's objectives.

    4. **Monitoring/Controlling**: Those processes performed to track, review, and regulate the progress and performance of the project and identify any areas in which changes to the Project Management Plan are required, and initiate any related changes.

    5. **Closing**: Those processes performed to finalize all work activities in order to formally close the project or Phase.

Project Manager: An organizational employee, representative, or consultant appointed to prepare project plans and organize the resources required to complete a project, prior to, during, and upon closure of the project life cycle. (See also *Project Management, Project Life Cycle.*)

Project Phase: A collection of logically related groupings of project activities that culminates in the completion of one or more deliverables. It is unique to a portion of the project or a major deliverable. Most or all five Process Groups may be executed in each. Generally completed sequentially, but may overlap, too. Phase characteristics include:

- Distinct focus of work from other Phases; organizations, location, or skill sets may differ.
- The Phase objective requires unique controls or processes for that Phase.
- All five Process Groups may be executed to provide added control and boundaries.
- Phase closure includes transfer or hand-off of the deliverable, i.e., tollgate, milestone, Phase review. Requires approval in most cases.
- May be sequential, overlapping, or predictive (fully plan driven with overlapping Phases).
- Phase examples: concept development, feasibility study, planning, design, develop prototype, build, test, train, and deploy.

Project Portfolio: A centralized collection of independent projects or programs that are grouped together to facilitate their prioritization, effective management, and resource optimization in order to meet strategic organizational objectives. (See also *Portfolio.*)

Project Portfolio Management (PPM): The combined coordination and management of all the active projects and programs to maximize the value they add to the organization by continuously monitoring their progress, prioritizing work, and allocating resources. It refers to the combined activity of all the active portfolios and independent projects going on within the organization and the improvement of the organization's capability, linking project/program/portfolio management with organizational facilitators (structural, cultural, technological, human resources practices) to support strategic goals. To apply this methodology, organizations must first measure their current capabilities, identify the future target capabilities, and then plan and implement improvements to close the gap between the two. (See also *Organizational Portfolio Management (OPM)*.)

Project Portfolio Management (PPM) System: An automated or electronic application that enables the centralized oversight of processes, methods, and technologies by Portfolio Leaders, Program Managers, Project Managers, and Project Management Offices (PMOs) to concurrently analyze and manage all proposed and active projects. (See also *Organizational Portfolio Management (OPM)*.)

Project/Program Manager and Portfolio Leader Competencies: Includes knowledge, performance, personal effectiveness. PMs focus on specific project objectives, control assigned project resources, and manage constraints.

Project Sponsor: A manager or executive internal to or outside of the Performing Organization who issues the Project Charter, ensures the alignment of the project with the strategic plan, and provides the resources, in cash or in kind, to undertake the project. (S)he is generally accountable for the development and maintenance of the project's Business Case document that quantifies the project's expected value and/or benefit to the Sponsoring or Requesting Organization. (See also *Initiating Sponsor and Sustaining Sponsor*)

Project Success: Completing the project within constraints of scope, time, cost, quality, resources, and risk (as approved by PMs and senior management) based on the last Project Performance Baseline approved (includes the Scope, Schedule, and Budget Performance Baselines).

Project Team: Includes the Project Manager and members of a team of people responsible for completing the project successfully. Roles include Analysts, Leads, Back-Ups, SMEs, Contractors, Business Partners, etc.

Receiving Organization: That entity which is the recipient of the product, service, system, or result of the project as delivered by the Performing Organization. (See also *Requesting Organization*.)

Requesting Organization: The entity that identifies an opportunity for improvement or a business problem, or makes a business request that establishes a set of expectations. It should then result in a set of requirements or scope of work that is included in a Request for Proposal, Request for Quotation, Request for Information, or an Invitation for Bids to one or more "Prospective Sellers." The "Selected Seller" will become the "Performing Organization" committed to producing a deliverable that will exploit the opportunity for improvement, solve the problem, or fulfill the request. Also known as the "Buyer," "Customer," "Client," "Investor," or "End-User."

Requirements: Functional or technical characteristics of a new product, service, system, or process that will meet a Requesting Organization's expectations or fulfill its request.

Risk Management Plan: A component of the Project Management Plan that describes how risk management activities will be structured and performed.

Risk Register: A document in which the iterative results of the risk identification, risk analysis, and risk response planning processes are recorded.

Rolling Wave Planning: An iterative approach in which a general macro plan is created for the current Phase based on what is known, then creating a more detailed plan for future Phases as more data is revealed or obtained.

SaaS (Software-as-a-Service): A licensing and delivery model where software is licensed, centrally hosted, and made available as a service, typically on a subscription basis. It moves the task of managing software and its deployment to third-party services.

Scope Creep: The insidious expansion of the product scope (i.e., expected functionality of the product, service, system, or result) that also expands or increases the project scope (i.e., the amount of work required) without a commensurate adjustment to or trade-off in the schedule, budget, resources, and/or quality, resulting in increased risk for the Performing Organization.

Sponsor: A manager or executive internal to or outside of the Performing Organization who issues the Project Charter, ensures the alignment of the project with the strategic plan, and provides the resources, in cash or in kind, to undertake the project. (S)he is generally accountable for the development and maintenance of the project's Business Case document that quantifies the project's expected value and/or benefit to the Sponsoring or Requesting Organization. (See also *Initiating Sponsor* and *Sustaining Sponsor.*)

Stakeholder: An individual, group, or organization that may affect, be affected by, or perceive itself to be affected by a decision, activity, or outcome of a project. The Project Team and any "interested entities" internal or external: sponsor, customers and users, sellers, business partners, organizational groups, functional managers, other (financial, government, SMEs, consultants).

Stakeholder Management Plan: A component of the Project Management Plan that describes how stakeholder management and engagement activities will be structured and performed.

Stakeholder Register: A project document that includes but is not limited to identification and assessment information, and the classification of the project stakeholders.

Strategic Business Plan: This plan focuses on what the organization is going to do to grow its market share. It is designed to answer the questions: What do we do? How can we beat the competition? It is directed at the product.

Strategy: It defines the way the mission will be accomplished. Using a well-defined strategy provides management with a thought pattern that helps it better utilize equipment and direct resources toward achieving specific goals. (For example, "the company will identify new customer markets within the United States and concentrate on expanding markets in the Pacific Rim countries.")

Sub-Portfolio: A smaller portion of a portfolio. For example, an organization's R&D activities may be directed at a number of different product types and, in each of the product types, a number of different projects are being actively pursued as a sub-portfolio. (See also *Portfolio.*)

Sub-Portfolio Leader: In some cases, due to the complexity and number of individual projects going on in a specific area, a major portfolio may be divided into sub-portfolios with a Portfolio Leader assigned to each of them. In these cases, each specific type of product may have a Portfolio Leader coordinating/managing all the individual projects/programs. Some organizations classify these individuals as Sub-Portfolio Leaders related to that product type. To reduce confusion, we use the term *Portfolio Leader* to represent both the Portfolio Leader and the Sub-Portfolio Leader.

Sustaining Sponsor: The individuals/group who have the political, logistical, and economic power to take over for the Initiating Sponsor to support the performance improvement project after the Align Stage has been completed all the way through to the Confirm Stage to sustain the gains achieved. (See also *Project Sponsor* and *Initiating Sponsor.*)

Sustain-the-Change Management Plan: A Key Output from the Confirm Stage in the contemporary framework that provides structure to guide the post-project Implementation Plan describing how to sustain the gains achieved during the performance improvement project that were approved by the Steering Committee.

Tailoring: The adjustment or refinement of a particular project, process (e.g., input, work activity, output), or Phase to meet the unique demands of a particular organization, environment, customer, or other stakeholder.

Target: The individual/group who must actually change as a result of the performance improvement project outcome(s).

Technology Management Plan: A subsidiary plan or component of the overall Project Management Plan, it should describe which project technology (both hardware and software) will be used to complete the project and how it will be implemented, documented, trained for, reported upon, and integrated into the daily work routines of the members of the performance improvement team.

Values: The basic beliefs or principles upon which the organization is founded and that make up its organizational culture. They are identified by top management and are rarely changed because they must be statements that the stakeholders hold and depend on as being sacred to that organization.

Value Proposition: A short statement (pre-project) that describes the tangible results or value a decision maker can expect from implementing a recommended course of action and its resulting benefit to the organization. It is expressed in a quantified fashion in the Business Case, where Value = Benefits − Cost (where Cost includes Risk). (See *Business Case.*)

Vision Statement: It provides a view of the future desired state or condition of an organization. (A vision should stretch the organization to become the best that it can be.) The Vision Statement provides an effective tool to help develop objectives.

Work Breakdown Structure (WBS): A hierarchical decomposition of the total scope of work to be carried out by the project team to accomplish the project objectives and create the required deliverables.

REFERENCES

1. Drawn from a blend of a 2010 *Quality Progress* online magazine article (http://asq.org/quality-progress/2010/02/qp-inbox.html); an MCTS blog, *The Continual Improvement vs. Continuous Improvement Dilemma* (http://www.mcts.com/Continual-vs-Continuous.htm); and the *PMBOK® Guide*, 6th Edition, p. 275, Project Management Institute, © 2017
2. Other definitions are drawn from those found in *Effective Portfolio Management Systems* by Voehl, Harrington, and Ruggles, © 2016, Boca Raton, FL: CRC Press, Appendix A: Project and OPM Definitions on pp. 135–145; *A Guide to the Project Management Body of Knowledge* (aka *PMBOK® Guide*), 6th Edition, © 2017, Project Management Institute, the Glossary on pp. 695–726; and The *Standard for Program Management*, 4th Edition, © 2017, Project Management Institute, pp. 50–52.

INDEX